Oxford **Mathematics** 3
Primary Years Programme

Contents

UNIT 1: TOPIC 1
Place value

Practice

1 Write these numbers in words.

a 6407 _____

b 5030 _____

c 9619 _____

2 Write the numerals for these numbers.

a Three thousand, five hundred and seventy-two _____

b One thousand, six hundred and five _____

c Eight thousand and eighty-four _____

3 Expand these numbers by place value.

a 2873 = _____ + _____ + _____ + _____

b 7091 = _____ + _____ + _____ + _____

c 4009 = _____

d 9797 = _____

4 Use place value to find the answers.

a 3000 + 600 + 50 + 2 = _____

b 8000 + 40 + 3 = _____

c 5000 + 300 + 7 = _____

d 100 + 8 + 6000 + 90 = _____

Which number is the largest? How do you know?

Challenge

1 Circle the value of the **red** digit.

a	4728	7 thousands	7 hundreds	7 tens	7 ones
b	9051	5 thousands	5 hundreds	5 tens	5 ones
c	1384	3 thousands	3 hundreds	3 tens	3 ones
d	4603	4 thousands	4 hundreds	4 tens	4 ones

2 Write the value of the **red** digit.

a 7580 _____   **b** 2095 _____

c 9143 _____   **d** 7851 _____

e 7194 _____   **f** 1902 _____

3 Write the numbers from question 2 in order from smallest to largest.

_____

4 Place the missing digits on the place value charts so that the numbers are in order from smallest to largest.

a

	Th	H	T	O
1	☐	9	4	8
3	1	☐	9	8
7	☐	5	2	0
8	6	☐	9	6
9	6	8	☐	1

b

	Th	H	T	O
0	4	☐	2	9
1	☐	1	2	8
3	4	1	☐	4
4	4	9	5	☐
5	☐	3	0	7

1 675 can be renamed as:

600 + 70 + 5, 670 + 5 and 600 + 75.

List as many ways to rename 8329 as you can think of.

2

a The combination code for Zephyra's bicycle lock has the digits 2, 3, 4 and 5 in it. List as many different possibilities of what it might be as you can.

b If the combination for Zephyra's bicycle lock is larger than 3200 but smaller than 3400, what might it be? _____ _____

c If the digit in the tens place of Zephyra's combination is larger than the digit in the ones place, what is it? _____

3 Will and Max both have house numbers in the thousands. Will's house number has a 2 in the hundreds column and Max's has a 7 in the tens column. If Will's house number is larger, what might their house numbers be? List four options.

Will	Max	Will	Max
_____	_____	_____	_____

Will	Max	Will	Max
_____	_____	_____	_____

OXFORD UNIVERSITY

Practice

1 Odd or even?

a 437 _____

b 964 _____

c 605 _____

d 8323 _____

e 2051 _____

f 3716 _____

g 9002 _____

h 2009 _____

i 7588 _____

2 Complete the number patterns and write *odd* or *even* beneath each number.

a Add 3

373	376	379	382					

odd _____ _____ _____ _____ _____ _____ _____ _____

b Add 5

1000	1005			1020	1025	1030		

_____ _____ _____ _____ _____ _____ _____ _____ _____

c Take away 2, add 5

20	18	23	21	26	24	29		

_____ _____ _____ _____ _____ _____ _____ _____ _____

3 Add 7 to each number and write whether the answer is odd or even.

Number	+7	Odd or even?
74		
391		
2150		
6708		
7463		

What happens when you add 7 to an odd number? What about to an even number?

1 Use the table below to predict whether the answer to each problem will be odd or even. Then use a calculator to find the answers.

What happens when you take an odd number away from an odd number?

Operation	Answer
even + even	even
even + odd	odd
odd + odd	even
odd + even	odd

a Rachael has $397 in one account and $2465 in a second account. How much money does she have in total?

Prediction: | odd | even | Answer: _____

b Ellen travelled 6014 km on Saturday and 2823 km on Sunday. How far did she go in total?

Prediction: | odd | even | Answer: _____

c A crowd of 1859 people attended the boat show on Saturday, while 3116 went on Sunday. How many people attended altogether?

Prediction: | odd | even | Answer: _____

d One cup of rice has around 6018 grains in it. How many grains would there be in 2 cups of rice?

Prediction: | odd | even | Answer: _____

e 4350 people lived in North Carson and 3879 lived in South Carson. How many people lived in Carson altogether?

Prediction: | odd | even | Answer: _____

1 London's game score had the digits 4, 9, 6 and 5 in it. If his score was even, what might it have been? List as many options as you can.

2 Matilda had 143 marbles. She bought between 250 and 350 more. How many might she have bought if she ended up with an even number? Record six different options.

_____ _____ _____ _____ _____ _____

3 Using the digits 1, 2, 8 and 7, make a thousands number that is:

a odd and larger than 7000. _____

b odd and smaller than 7000. _____

c even and larger than 7000. _____

d even with a 2 in the hundreds column. _____

What is the largest 4-digit odd number you can make using any digits?

4 Use the digits 1, 7, 2, 9 and 6 to make:

a the largest odd number possible. _____

b the largest even number possible. _____

c the smallest odd number possible. _____

d the smallest even number possible. _____

Practice

1 A Year 3 class learned four different mental addition strategies: extending number facts, extending doubles facts, the split strategy and rearranging numbers to make them easier to add. Record the strategy they used to solve each problem.

a $16 + 7 + 13 + 4$

$= 16 + 4 + 7 + 13$

$= 20 + 20 = 40$

_____

b $43 + 6$

I know $3 + 6 = 9$,

so $43 + 6 = 49$

_____

c If $10 + 10 = 20$,

then $100 + 100 = 200$

_____

d $38 + 41$

$= 30 + 40 + 8 + 1$

$= 70 + 9 = 79$

_____

2 Use a mental strategy to solve the following addition problems. Record the strategy you used.

Strategy

a $12 + 4 + 8 + 16 =$ _____   _____

b $36 + 32 =$ _____   _____

c $40 + 40 =$ _____   _____

d $37 + 12 =$ _____   _____

OXFORD UNIVERSITY P

Challenge

1 Holiday Falls has a range of visitor activities. Look at the price list to find:

a two activities for which you could use the extended doubles strategy to add the prices.

_____

_____

Holiday Falls price list

Bungee jumping	$250
Horse riding	$98
Jetboating	$143
Zip lining	$152
Helicopter ride	$250
City tour – full day	$137
City tour – half day	$88
Dolphin cruise	$102

b how much your chosen activities cost altogether. _____

2 Choose four activities that you could add the prices of by using the 'rearrange the numbers' strategy.

a List the activities and their prices.

_____ _____ _____ _____

b Add the prices in your head and record the total. _____

3 Choose two activities that will cost less than $300 together.

a List the activities and their prices.

_____ _____

b Add the prices using the split strategy. _____

4 Use a mental strategy of your choice to add the prices of:

a zip lining and a dolphin cruise. _____

b jetboating, horse riding and a full-day city tour. _____

c the three most expensive activities. _____

d the four cheapest activities. _____

1 Together, Alia and Adnan had 94 trading cards. Use a mental addition strategy to work out how many cards they each might have had. Record at least five possibilities.

2 Dora loves to double numbers. Use a mental addition strategy to help her double:

a   32 _____

b   45 _____

c   27 _____

d   54 _____

e   48 _____

f   66 _____

3 Choose four different sets of four numbers from the box below that you think will be easy to add using a mental strategy. Record the numbers and use a strategy such as rounding to estimate the answers. Add the numbers in your head and record your answers, checking if they are close to your estimates.

		31		25
17	28	15	42	37
45		27		
19	26	23	33	14

		Estimate	Exact answer
a	_____ + _____ + _____ + _____ =		
b	_____ + _____ + _____ + _____ =		
c	_____ + _____ + _____ + _____ =		
d	_____ + _____ + _____ + _____ =		

OXFORD UNIVERSITY P

Practice

1 Use the jump strategy to solve each addition problem. Then check your answer using vertical addition.

a   37 + 42 = ☐

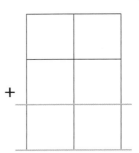

b   253 + 615 = ☐

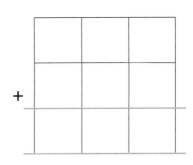

c   828 + 140 = ☐

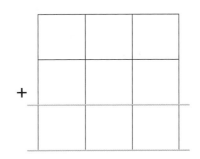

d   601 + 292 = ☐

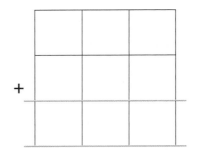

1 Use a written method of your choice to solve the following problems. Show how you found the answers.

a Sam sold 243 donuts at the Saturday market and 456 on Sunday. How many did she sell over the weekend?

b Ollie sold 705 pies on Saturday and 283 pies on Sunday. How many did he sell over the weekend?

2 Daria's chickens live in a field. This table shows how many eggs her chickens laid over three weeks.

Week 1	Week 2	Week 3
425	210	304

a How many eggs did Daria's chickens lay in total? Show your working.

Daria's total

Henry's chickens live in a barn. This table shows how many eggs they laid in three weeks.

Week 1	Week 2	Week 3
206	331	212

b How many eggs did Henry's chickens lay in total? Show your working.

Henry's total

c Whose chickens laid the most eggs in total? _____

d Whose chickens had laid the most eggs after two weeks? _____

OXFORD UNIVERSITY F

1   Tetsu and Tia's grandmother gave them $150 each. At the end of one year, they had a total of $647 between them. Use a written addition strategy to work out how much they might each have. Find at least five solutions.

2   Spencer had 958 flowers to deliver to three different shops. How many flowers might he have delivered to each shop? Show at least three solutions.

3   Using the digits 1, 2, 3, 4, 5 and 6, make two 3-digit numbers that total more than 700 but less than 900. Show your working.

4   Using the digits 2, 3, 4, 5, 6, 7, 8 and 9, make two 4-digit numbers that total less than 10 000. Show your working.

Practice

1　Solve each problem by subtracting to the nearest 10 on the number line and then completing the operation.

a　$57 - 8 = 57 - \boxed{} - \boxed{} = \boxed{}$

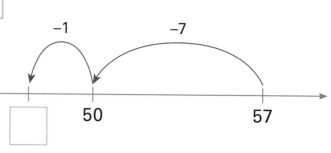

−1　　−7

$\boxed{}$　　50　　　　　57

b　$73 - 7 = 73 - \boxed{} - \boxed{} = \boxed{}$

c　$141 - 9 = \boxed{} - \boxed{} - \boxed{} = \boxed{}$

2　Write whether each problem is using the split strategy or extending number facts. Then solve the problems.

Strategy

a　$153 - 31 = 153 - 30 - 1 = \boxed{}$

_____

b　$456 - 115 = 456 - 100 - 10 - 5 = \boxed{}$

_____

c　If I know $7 - 4 = 3$, I know $57 - 4 = \boxed{}$

_____

d　$49 - 37 = 49 - 30 - 7 = \boxed{}$

_____

e　If I know $5 - 2 = 3$, I know $165 - 2 = \boxed{}$

_____

1 Southeast School held their annual concert over three nights.
The hall can fit an audience of 367 people.
Use the split strategy to show how many people attended if:

a 47 seats were empty.

_____

b 29 seats were empty.

_____

c 133 seats were empty.

_____

2 Here is the attendance for Southeast School's concert last year:

Night 1: 245   Night 2: 163   Night 3: 366

Use rounding to estimate the difference between the numbers of people on the nights listed. Then use your choice of mental strategy to find the exact answers.

a Night 1 and Night 2: Estimate _____ Exact answer _____

b Night 2 and Night 3: Estimate _____ Exact answer _____

c Night 1 and Night 3: Estimate _____ Exact answer _____

3 This table shows the number of students in each year level at Southeast School and how many were away from school the day after the concert.

Use the getting to a 10 strategy to work out how many students from each year level were at school.

Year level	Number of students	Number absent	Number present
F	37	9	
1	42	6	
2	34	8	
3	73	5	
4	66	9	
5	41	4	
6	45	7	

1 Farmer Betty had 174 peaches growing in her orchard. One night there was a big storm and some of the peaches were blown off the trees. How many peaches might have been left on the trees? Find at least five solutions.

2 Jensen the baker had 113 cupcakes left after two days of sales. How many might he have started with and how many might he have sold each day? Show at least three solutions.

3 Using the digits 3, 4, 5 and 6, make two 2-digit numbers that could each be subtracted from 100 to leave between 45 and 55. Show your working.

4 Using the digits 1, 2, 3, 4, 5, 6 and 7, make two 3-digit numbers that could each be subtracted from 500 to leave less than 200. Show your working.

Practice

1 Use the jump strategy to solve each subtraction problem. Then check your answer using vertical subtraction.

a   89 – 36 = ☐

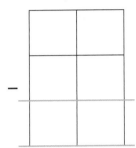

b   458 – 35 = ☐

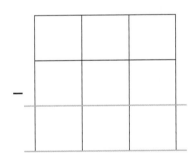

c   634 – 231 = ☐

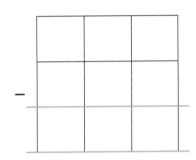

d   842 – 320 = ☐

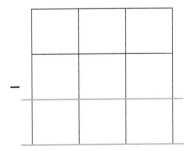

Challenge

1 Use a written method of your choice to solve the following.
Show how you found the answers.

a Laney had $758. After she bought a ticket to a concert, she had $346 left. How much did the ticket cost?

b Jenn had $673. She bought clothes for $263. How much money did she have left?

2 Matthias loves to read. Here's how many pages he read in a month.

Week 1	Week 2	Week 3	Week 4
312	898	514	451

Show your working as you find how many more pages he read:

a in week 2 than in week 1.

b in week 3 than in week 1.

c in week 3 than in week 4.

d in week 2 than in week 3.

1 Daisy had 2574 glasses in her restaurant. A shelf fell down and a lot of the glasses broke. If she had between 1600 and 1700 glasses left, how many glasses might have smashed? Show at least five solutions.

2 Kai baked 836 cookies. He sold some on Saturday and some on Sunday, leaving him with 97. How many cookies might he have sold on each day? Show at least three solutions.

3 There were 457 people left in the stadium half an hour after the game. How many people might have attended, and how many have already left? Show at least three solutions.

Practice

1 Write two subtraction facts for each addition fact.

a $26 + 34 = 60$

b $58 + 67 = 125$

_____

_____

_____

_____

c $162 + 309 = 471$

d $853 + 145 = 998$

_____

_____

_____

_____

2 Write two addition facts for each subtraction fact.

a $67 - 49 = 18$

b $93 - 28 = 65$

_____

_____

_____

_____

c $386 - 97 = 289$

d $804 - 673 = 131$

_____

_____

_____

_____

3 Use the following numbers to make two addition and two subtraction facts:

529, 877, 348

_____

_____

_____

_____

Challenge

1 Write a number sentence to solve the following problems. Then write an inverse operations number sentence to check your answer.

a Jimmy had 132 beads. He used 71 to make a necklace. How many does he have left?

b Josh has five boxes with nine chocolates in each. How many chocolates does he have altogether?

c Charlie has 257 erasers in his collection. Keira has 124. How many do they have altogether?

d Erin shares 36 sweets equally into nine bags. How many does she put in each bag?

---

2 Use the compensation (rounding) strategy to solve the following. Show your working.

a Emma ran 46 km one day and 29 km the next. How far did she run altogether?

b Kwabi had 323 books. He bought another 38. How many does he have now?

1 Nam got his inverse operations homework caught in the zip of his schoolbag and some of the numbers got ripped. What might the missing numbers be? Record at least three solutions.

a ☐ 6 + 3 ☐ = ☐ ☐

b ☐ 8 ☐ − ☐ 6 = ☐ ☐ ☐

3 ☐ + ☐ 6 = ☐ ☐

☐ 8 ☐ − ☐ ☐ ☐ = ☐ 6

☐ ☐ − ☐ 6 = 3 ☐

☐ 6 + ☐ ☐ ☐ = ☐ 8 ☐

☐ ☐ − 3 ☐ = ☐ 6

☐ ☐ ☐ + ☐ 6 = ☐ 8 ☐

☐

☐

2 Kaylee added 30 to a number, then subtracted 17. What number might she have started with and what number does she end with? Find at least three solutions.

☐

3 Lincoln multiplied a number by two, and then added 21. What number might he have started with and what number does he end with? Find at least three solutions.

☐

Practice

1 Write two division facts for each multiplication fact.

a $7 \times 8 = 56$

b $9 \times 4 = 36$

_____

_____

_____

_____

c $100 \times 5 = 500$

d $7 \times 12 = 84$

_____

_____

_____

_____

2 Write two multiplication facts for each division fact.

a $28 \div 7 = 4$

b $54 \div 9 = 6$

_____

_____

_____

_____

c $99 \div 11 = 9$

d $400 \div 100 = 4$

_____

_____

_____

_____

3 Use the following numbers to make two multiplication and two division facts:

10, 250, 25

_____

_____

_____

_____

1 Write a number sentence to solve each of these problems. Then write an inverse operations number sentence to check your answer.

a The school choir had five rows with eight students in each row. How many students were in the choir?

b Marshall arranged his coffee mugs in six rows of seven. How many mugs did he have altogether?

c Lois shared 70 pieces of popcorn between seven people. How many pieces did each get?

d Rosalie has 63 toy cars. She puts them into nine equal rows. How many are in each row?

2 Dylan jumbled up his inverse operations homework. Rewrite the fact families correctly.

a $52 \times 4 = 13$

$4 \div 13 = 52$

$13 \div 52 = 4$

$13 \times 52 = 4$

b $17 \times 255 = 15$

$255 \times 15 = 17$

$17 \div 255 = 15$

$15 \div 17 = 255$

1 Make as many multiplication and division facts as you can using any of the following numbers:

14   32   12   36   4   6   9   28   7   8   3   2

2 Conroy made a big rubber band ball using between 29 and 100 rubber bands. If the number of rubber bands could be shared equally into both three and five groups, how many might he have used?

3 Taneisha multiplied two numbers together to get 48. What might the numbers have been? Find at least three solutions.

Practice

1 Chidi used three different strategies in his mathematics homework. Draw lines to match the strategy he used to solve each problem.

a
$$32 \div 4 = 32 \div 2 \div 2$$
$$= 16 \div 2$$
$$= 8$$

skip counting

double and double again

halve and halve again

b | $6 \times 4 = 4, 8, 12, 16, 20, 24$

c
$24 \div 3$ Think $3 \times 8 = 24$
So $24 \div 3 = 8$

using multiplication facts for division

d
$$4 \times 9 = 2 \times 9 + 2 \times 9$$
$$= 18 + 18$$
$$= 36$$

Would you have used the same strategies as Chidi?

2 Use a mental strategy of your choice to solve the following. Record the strategy you used.

Strategy

a $7 \times 3 =$ _____   _____

b $10 \times 4 =$ _____   _____

c $4 \times 25 =$ _____   _____

d $36 \div 4 =$ _____   _____

e $45 \div 5 =$ _____   _____

f $4 \times 30 =$ _____   _____

OXFORD UNIVERSITY PR

1 There are four people in the Hendricks family and they love doing things together. Use a mental strategy to work out the following problems:

a It costs $16 per person to go ice skating. How much would it cost the Hendricks family to go?

_____

b Dad bought a family-sized box of crackers. If there were 80 crackers in the box and they divided them equally, how many would each person in the Hendricks family get?

_____

c A movie ticket costs $15 for an adult and $9 for a child. How much would two of each ticket cost?

_____

d The family went on a bike-riding holiday. If each person rode the same distance, and the family rode a total of 320 km altogether, how far did each person ride?

_____

e Each family member had a budget of $40 to buy gifts for each other. How much did they have to spend in total?

_____

2 Julian discovered that to multiply a number by 3, you can double it and add the number again: $8 \times 3 = 8 \times 2 + 8 = 16 + 8 = 24$

Try Julian's strategy to find:

a $11 \times 3 =$ _____

b $23 \times 3 =$ _____

c $70 \times 3 =$ _____

d $120 \times 3 =$ _____

e $36 \times 3 =$ _____

1 Freya filled a container with free samples of chocolate at the chocolate factory. The number of pieces she had can be shared equally between three. How many pieces might be in the container? Find at least five solutions.

2 Jayda saved four times as much money as her little brother. How much money might they both have? Show at least five solutions.

3 Li is planning a party with five guests, including himself. Help by showing:

a   what food he is serving.

b   how many of each item his guests will have.

c   how many of each item he needs to buy in total.

Choose at least three different food options for Li's party.

OXFORD UNIVERSITY P

Practice

1 Use the split strategy to solve each problem. Then check your answers using the grid method.

a 4×37 = ☐ × ☐ + ☐ × ☐

= ☐ + ☐

= ☐

×		

= ☐

b 5×64 = ☐ × ☐ + ☐ × ☐

= ☐ + ☐

= ☐

×		

= ☐

c 8×55 = _____

= _____

= _____

×		

= ☐

d 9×32 = _____

= _____

= _____

×		

= ☐

e 6×46 = _____

= _____

= _____

×		

= ☐

Challenge

1 Use a written method of your choice to solve these problems. Show how you found the answers.

a Eden has seven cartons of eggs with 24 eggs in each. How many eggs does she have altogether?

b Marco made 78 muffins each day from Monday to Friday. How many did he make altogether?

---

2 This table shows how far different teachers travel to get to and from school each day.

Use a written method to work out:

Teacher	Distance
Mr Smith	53 km
Mrs Burton	71 km
Miss White	22 km
Mrs Swain	48 km
Mr Chappel	39 km

a how far each teacher travels in a school week.

b how far Mrs Burton travels in two school weeks.

c how far Mr Chappel travels in four school days.

Mastery

1 Cara was trying to find the best way to package sweets in her factory. The packing machine can be set to make five, six or seven packs at a time. Each pack can have 28, 37 or 56 sweets in it. Use written multiplication to find a packaging method that will use:

a close to 200 sweets.

b more than 340 sweets.

c between 200 and 300 sweets.

d exactly 185 sweets.

2 Using any of the digits 5, 6, 7, 8 and 9, make a one-digit and a two-digit number that equal between 350 and 450 when multiplied together. Show your working.

3 Using any of the digits in question 2, make a one-digit and a two-digit number that equal more than 500 when multiplied together. Show your working.

Practice

1 Find the answers. Then rearrange the numbers and add or multiply them in a different order to check.

a $14 + 9 + 6 =$ _____   _____

b $32 + 7 + 8 + 23 =$ _____   _____

c $15 + 24 + 16 + 25 =$ _____   _____

d $7 \times 2 \times 3 =$ _____   _____

e $2 \times 9 \times 5 =$ _____   _____

f $6 \times 5 \times 4 \times 2 =$ _____   _____

2 Rotate and redraw each array and write the new multiplication fact.

a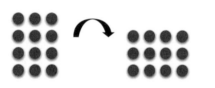

$4 \times 3 = 12$ ____ × ____ = ____

b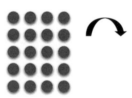

$5 \times 4 = 20$ ____ × ____ = ____

c

$8 \times 3 =$ ____ _____

d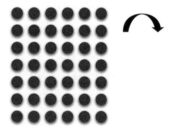

$7 \times 6 =$ ____ _____

OXFORD UNIVERSITY

Challenge

1 Use your knowledge of "undoing" the problem to solve the following.

a Bella thought of a number. She divided it by 2 and got 15. What was her number? _____

b Sia thought of a number. She added 12 and got 57. What was her number? _____

c Chris thought of a number. He added 2 and then multiplied it by 2 and got 20. What was his number? _____

d Ivan thought of a number. He subtracted 10 and then divided it by 3 and got 8. What was his number? _____

---

2 Add +, −, × or ÷ to make the pairs of equations true. Then write two other linked facts using the same numbers.

a 19 ☐ 23 = 42   42 ☐ 23 = 19

_____   _____

b 8 ☐ 9 = 72   72 ☐ 8 = 9

_____   _____

> Linked number facts are also called fact families because they are related.

c 70 ☐ 10 = 7   7 ☐ 10 = 70

_____   _____

d 84 ☐ 36 = 48   48 ☐ 36 = 84

_____   _____

e 4 ☐ 8 ☐ 3 = 96   96 ☐ 3 ☐ 8 = 4

_____   _____

1   Usman has 4 fish tanks, each with a different number of fish in it. Two of the numbers added together make 50, while the other two make 75. How many fish might there have been in each tank and how many fish does Usman have in total? Show at least three solutions.

2   Alice arranged her chocolates in 3 boxes with 2 layers of 8 chocolates in each. She recorded this as 3 × 2 × 8 = 48. How else could she have organised her 48 chocolates in boxes with equal numbers in each? Draw and write as many solutions as you can.

Practice

1 Shade the following fractions.

a $\frac{2}{5}$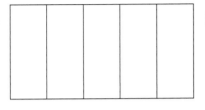

b $\frac{3}{8}$

c $\frac{2}{3}$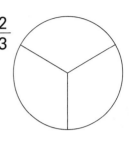

d $\frac{3}{4}$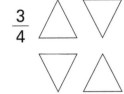

e $\frac{4}{5}$

f $\frac{7}{8}$

2 Divide each shape into the fractions shown.

a

quarters

b

thirds

c

sixths

d

halves

e

eighths

f

quarters

3 Colour two parts of each of the fractions in question 2.

4 Write the name of each fraction you have coloured.

a _____

b _____

c _____

d _____

e _____

f _____

Challenge

1

a   Colour the circles in three different colours.

b   Record what fraction of each colour there is.

_____

2

a   Colour the circles in four different colours.

b   Record what fraction of each colour there is. _____

3   Redraw and divide the rectangle into thirds in as many different ways as you can.

4   Redraw and divide the rectangle into quarters in as many different ways as you can.

1 Gianni gave away $\frac{3}{5}$ of his crackers.

a Draw a diagram to show what this might look like.

b How many crackers might Gianni have started with? _____

c How many did he give away?

_____

d How many does he have left?

_____

e What fraction does he have left? _____

2 Eleanor had $\frac{1}{4}$ of a packet of sweets left.

a Draw a diagram to show what this might look like.

b How many sweets might Eleanor have started with? _____

c How many has she eaten?

_____

d How many does she have left?

_____

e What fraction did she eat? _____

3 Jonas and Sarah shared a cake that was divided into six pieces. Draw and label at least three options showing what fraction they might each have eaten.

Practice

1 Label the fractions on each number line.

a

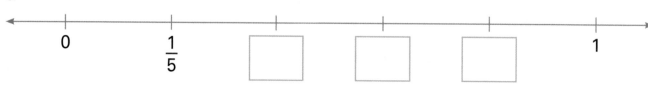

b

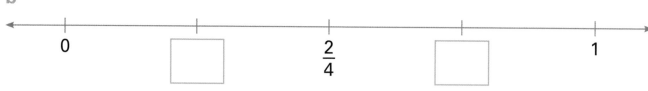

c

d

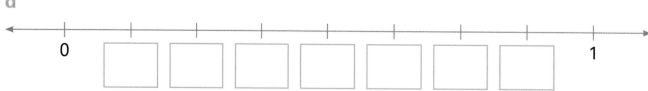

e

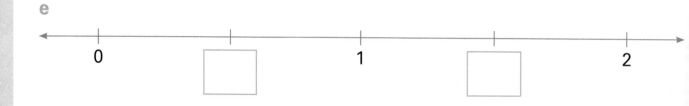

OXFORD UNIVERSITY P

Challenge

1 Make a number line to show:

a counting by eighths from $\frac{1}{8}$ to $\frac{6}{8}$.

b counting by fifths from $\frac{3}{5}$ to $1\frac{3}{5}$.

c counting by thirds from 0 to $\frac{6}{3}$.

d counting by halves from 0 to $3\frac{1}{2}$.

e counting by quarters from 0 to $2\frac{3}{4}$.

2 How many:

a fifths in $1\frac{3}{5}$?

b halves in $3\frac{1}{2}$?

c quarters in $2\frac{3}{4}$?

d thirds in 2?

RD UNIVERSITY PRESS

1

a   Show counting from 0 to 2 by quarters and by eighths on the same number line.

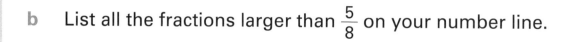

b   List all the fractions larger than $\frac{5}{8}$ on your number line.

_____

c   List all the fractions smaller than $\frac{2}{4}$ on your number line.

_____

d   List the pairs of fractions that are the same size on your number line.

_____

2   Three children started a 1 km run at the same time. Show what fraction of the distance each has travelled on the number lines.

a   Lando has run $\frac{2}{3}$ of the course.

0   1

b   Archer has run $\frac{1}{4}$ of the course.

0   1

c   Nala has run $\frac{4}{5}$ of the course.

0   1

Use your number lines to decide:

d   who has travelled the furthest. _____

e   who still has the longest distance to run. _____

OXFORD UNIVERSITY

Practice

1 Draw three different ways you could make $1 with coins.

2 Draw three different ways you could make 90c with coins.

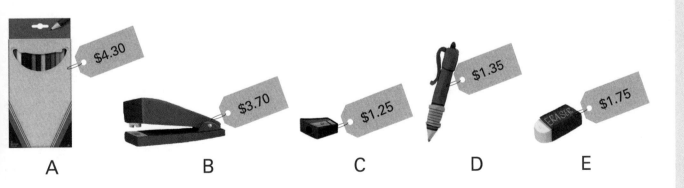

3 Write the letter of the item that would give you:

a   change of 25c from $2. ☐

b   change of 70c from $5. ☐

c   change of 30c from $4. ☐

d   change of $1.75 from $3. ☐

A $4.30

B $3.70

C $1.25

D $1.35

E $1.75

ORD UNIVERSITY PRESS

Challenge

The currency of Japan is the Yen. Here are the coins:

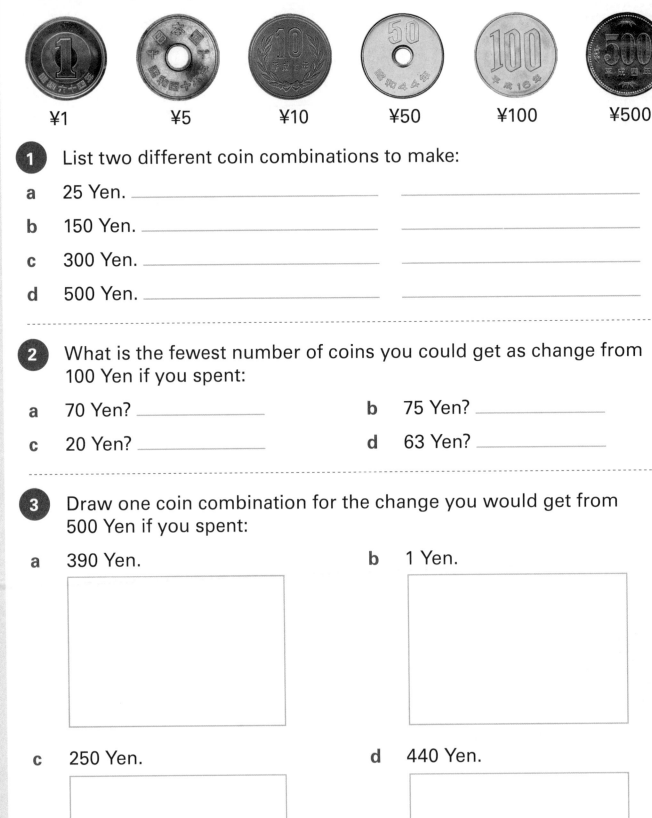

¥1   ¥5   ¥10   ¥50   ¥100   ¥500

1 List two different coin combinations to make:

a 25 Yen. _____   _____

b 150 Yen. _____   _____

c 300 Yen. _____   _____

d 500 Yen. _____   _____

- -

2 What is the fewest number of coins you could get as change from 100 Yen if you spent:

a 70 Yen? _____   **b** 75 Yen? _____

c 20 Yen? _____   **d** 63 Yen? _____

- -

3 Draw one coin combination for the change you would get from 500 Yen if you spent:

a 390 Yen.

b 1 Yen.

c 250 Yen.

d 440 Yen.

OXFORD UNIVERSITY P

1. Ming bought two items with a $10 note and received change of $4.20. How much might each of the items have cost? Show at least three solutions.

2. Eva had some notes and coins in her purse that totalled $10. Show at least five different combinations she might have had.

3. Ruby received $1.65 in change. How much might she have spent and what notes and coins might she have paid with? Find at least two solutions.

Practice

1 Write the rule and finish the pattern.

a   Rule: _____

70	65	60	55						

b   Rule: _____

57	59	61							

c   Rule: _____

17	24		38						

d   Rule: _____

	88	78	68						

2 Circle the numbers that would be part of this pattern if you continued it.

a

26	30	34	38	42	46

52	58	60	75	70

b

16	24	32	40	48	56

72	88	94	96	100

c

99	92	85	78	71	64

59	50	36	27	8

How can you tell if the rule involves subtraction?

OXFORD UNIVERSITY

The function machine is not working properly. Write the rule and correct the error for each one.

a Rule: _____

In	Out
43	32
86	75
31	20
45	35
78	67

b Rule: _____

In	Out
55	58
102	105
98	100
67	70
114	117

c Rule: _____

In	Out
150	160
245	254
303	313
99	109
225	235

d Rule: _____

In	Out
200	188
56	44
61	49
84	72
72	68

---

2 Milos organised some numbers into a group. He put the numbers that didn't fit his rule in the Out group.

In	57	15	91	143	73	29

Out	62	12	102	90	134	6

a Write the rule for the In group. _____

b Write some numbers that would fit into the In group.

_____

c Write some more numbers that would fit into the Out group.

_____

1 Heidi counted by 4 from a number to find the next six numbers. What might her start number have been, and what were the other numbers? Record at least three possibilities.

2 Ethan used a two-step rule to get from 1 to 16. What might the rule have been? Show at least two possibilities.

3 Deanna made a two-step pattern. Use words, symbols or a table to represent what her pattern might have been if her starting number was 10.

OXFORD UNIVERSITY

Practice

1 Write a number sentence to solve the word problems.

a Lindy ate 47 grams of chocolate on Monday and the same amount on Tuesday. How much chocolate did she eat altogether?

b Grandpa had $143. He gave $37 to Jaydyn and $24 to Jasmine. How much money did he have left?

c Megan took 213 photos in the first week of her holiday and 157 in the second week. How many photos did she take in total?

d Van had 56 trading cards. He bought 28 more and sold 40 of them. How many did he end up with?

2 Use – or + and an equals sign to balance each equation.

a 13 ☐ 11 ☐ 24

b 10 ☐ 18 ☐ 8

c 27 ☐ 14 ☐ 13

d 42 ☐ 23 ☐ 19

Challenge

1 Circle the correct number to balance the equations.

a $3 + 21 = 30 - \boxed{}$

| 24 | 16 | 6 | 5 |

b $48 - 30 = \boxed{} + 11$

| 7 | 14 | 18 | 1 |

c $67 - \boxed{} = 93 - 49$

| 3 | 44 | 32 | 23 |

d $\boxed{} + 32 = 23 + 20$

| 43 | 11 | 75 | 17 |

2 Students at Equation School voted on a school pet.
Here are the results:

Pet	Number of votes
Cat	204
Lizard	96
Stick insect	32
Parrot	113
Fish	28

How could you check your answers?

a How many people voted for the two most popular pets? _____

b How many people voted for the two least popular pets? _____

c How many more votes did lizards get than stick insects? _____

d Which got more votes, cats by themselves or lizards, stick insects

and parrots together? _____

e Which two animals got between 220 and 240 votes together?

_____

f Which animal got 81 votes more than stick insects? _____

OXFORD UNIVERSITY

1 Use + or – to make the equations balance.

a   7 ☐ 6 ☐ 7 = 20

b   25 ☐ 6 ☐ 2 = 21

c   25 ☐ 10 ☐ 4 = 11

d   58 ☐ 12 ☐ 4 = 74

e   105 ☐ 7 ☐ 8 ☐ 12 = 116

f   132 ☐ 5 ☐ 6 ☐ 3 = 124

2 Between them, Anton, Jay and Mia picked 210 oranges. How many might they each have picked? Show at least three solutions.

3 Maddie made 288 lollipops. She packed them into bags each with the same number of lollipops. How many bags might she have used and how many lollipops were in each? Find as many solutions as you can.

Practice

1 Match each item with its most likely length in real life.

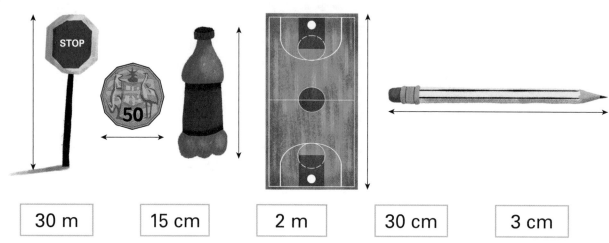

30 m	15 cm	2 m	30 cm	3 cm

2 **a** Draw four different shapes, each with an area of 10 cm².

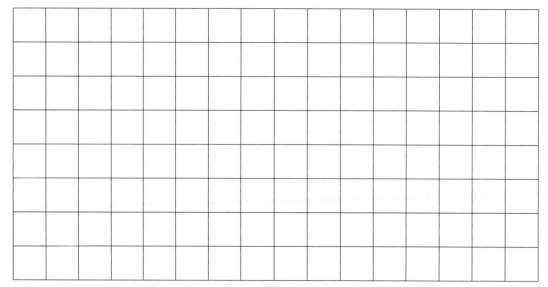

b What is the total area of the shapes you drew? _____

c What area of the grid is not taken up by your shapes? _____

d Choose two of your shapes and record the length of each of their sides in centimetres.

Shape 1: _____ Shape 2: _____

OXFORD UNIVERSITY

Challenge

1 Measure each of the lines in cm (to the nearest centimetre) and in mm.

_____

_____ cm or _____ mm   _____ cm or _____ mm

2 Draw a line exactly:

a   6 cm long.

b   44 mm long.

c   3 cm long.

d   3 mm long.

3 Find the area of each shape.

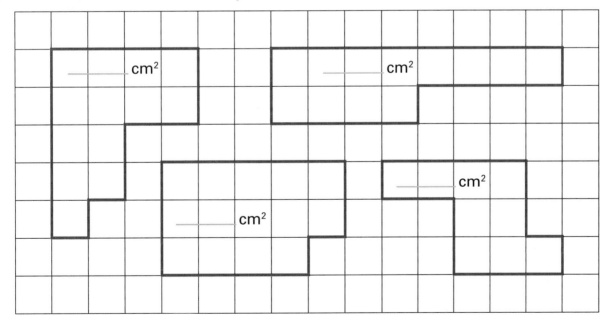

_____ cm²   _____ cm²

_____ cm²

_____ cm²

1 Draw a picture of a house and garden with:

a a door that is 2 cm high.

b a pet that is 1 cm high.

c a fence that is 10 cm long.

d a sun that is 25 mm wide.

2 Label the length of five other items in your picture.

3 Fred has red floor tiles that are 4 m square and blue tiles that are 3 m long and 2 m wide. Use some of both tiles to completely cover Fred's floor.

Practice

1

a Draw an object with three layers and six cubes in each layer and record its volume.

b Draw an object with two layers and eight cubes in each layer and record its volume.

Volume: _____

Volume: _____

2 Match the containers with their likely capacities.

a

b

c

9 litres

200 mL

5 mL

4 litres

500 mL

2 litres

d

e

f

Which of the containers has a capacity most similar to a drink bottle?

Challenge

1 Decide whether each story is about volume or capacity. Then solve the problem.

a Kayla had a full bottle of orange juice. She shared the drink equally into four cups. If the bottle held 2 litres, how much did each person get?

| volume | capacity |

Answer: _____

b Tariq's water tank had 120 litres in it. If he used 10 litres per day for a week, how much water would he have left?

| volume | capacity |

Answer: _____

c Nea wants to build a model building with four layers, and 8 cubic centimetres in each layer. How many cubic centimetres will she need?

| volume | capacity |

Answer: _____

2

a List five things that you think have a capacity greater than 1 litre.

b List five things that you think have a capacity of less than 1 litre.

OXFORD UNIVERSITY P

1 Hugh made an object with three layers and a volume of 17 cm³. What might it have looked like? Draw three different options.

2 Devash had three containers with a total capacity of 2 litres. What might the capacity of each container have been? Show three options.

3 Alana's four dogs each has a bowl with a different capacity. If the total capacity is 1600 mL, what might the capacity of each bowl be?

Practice

1 Sort the items into the correct categories by writing the letters in the table.

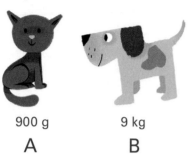

900 g
A

9 kg
B

Lighter than 1 kg	1 kg	Heavier than 1 kg

90 g
C

1000 g
D

148 g
E

10 g
F

600 g
G

100 g
H

2 Find two items from question 1 that:

a have a total mass of less than 1 kg. _____

b have a total mass of exactly 1 kg. _____

c have a total mass of more than 1.5 kg and less than 2 kg.

_____

d have a total mass of more than 2 kg. _____

3 Match the correct mass to each item.

| 7.5 kg |
| 75 g |
| 75 kg |
| 7.5 g |

a

b

c

d

_____ _____ _____

1 Choose pairs of objects and draw them on the correct ends of the balance scales. Use each object only once.

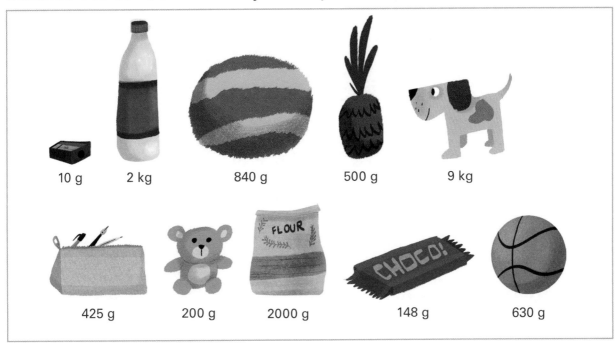

10 g   2 kg   840 g   500 g   9 kg

425 g   200 g   2000 g   148 g   630 g

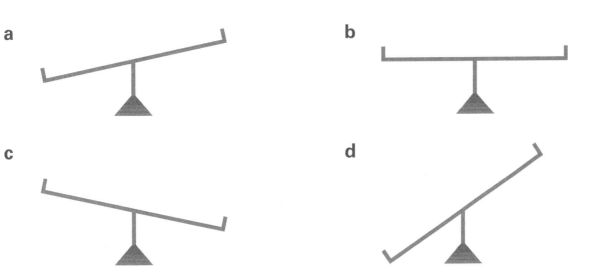

a

b

c

d

2 Read the scales and record the mass of each item.

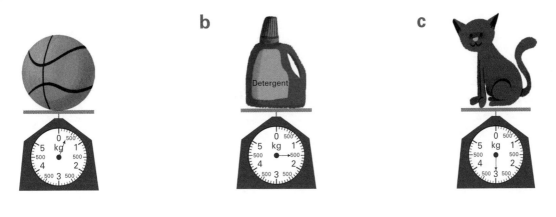

a

b

c

Detergent

_____ _____ _____

1 Zoe has three items that have a total mass of 3 kg. What might the items be and what might be the mass of each? Show at least three solutions.

2

a Draw six animals and estimate the mass of each.

b Order your animals from heaviest to lightest.

Practice

1   Circle the time that matches each clock.

a

10 past 2

10 to 2

2 past 10

b

6:20

4 past 6

20 to 6

c

1:08

8:01

5 past 8

d

8 to 1

20 to 1

1:40

e

8:45

9:45

9:42

f

2 minutes to 4

18 minutes past 11

2 minutes past 4

- -

2   Write each time in analogue and digital time.

a

_____

[:   ]

b

_____

[:   ]

c

_____

[:   ]

Challenge

1 It takes Tao 27 minutes to walk to school. Show on the clocks what time he left home if he arrived at:

a   8:30

b   8:14

c   8:42

2 Allegra gets driven to school. The clocks show when she left home and when she arrived at school today.

Left home

Arrived at school

a Write the time she left home in analogue time.

_____

b Write the time she arrived at school in digital time. ☐ : ☐

c How long did the trip take?

_____

3 Yusra went shopping at 10:30. Below, she has written how much later she did other activities. Mark the start time of each activity on the clocks.

a

Walk: 1 hour and 20 minutes later

b

Nap: 3 hours and 42 minutes later

c

TV: 6 hours and 50 minutes later

d

Dinner: 8 hours and 17 minutes later

e

Brush teeth: 11 hours and 49 minutes later

f

Bed: 12 hours later

1 Alec woke up at a time with the numbers 2, 3 and 5 in it. What time might it have been? Draw analogue clocks to show three different options and write the digital and analogue times for each.

2 Sebastian did each of the following activities in one day:

- went to the movies
- cooked lunch
- played basketball
- visited a friend
- read a book
- watched TV.

Draw clocks and write the analogue times to show when he might have done each activity through the day. Then record how long each activity might have taken.

Practice

1 Sort each shape as regular or irregular by writing its letter in the correct column of the table.

Regular	Irregular

A   B   C

D   E   F

2 Draw a shape that:

a is irregular.

b has one pair of parallel sides.

c has two pairs of parallel sides.

d has four right angles and is regular.

e has four right angles and is irregular.

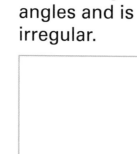

f has no right angles and is regular.

1 Draw and describe the following quadrilaterals (four-sided shapes):

a square

b rhombus

c trapezium

d kite

2 Choose two of the shapes in question 1 and compare them on a Venn diagram.

Shape 1: _____

Shape 2: _____

1 Make as many different shapes as you can by combining these shapes. Label each one.

2 Louisa drew a picture with six regular shapes and six irregular shapes. What might it have looked like?

Practice

1 Sort each 3D shape as prism, pyramid or other by writing its letter in the correct column of the table.

Prism	Pyramid	Other

A   B   C   D

E   F   G   H

2 Draw a 3D shape that has:

a at least two square faces.

b at least two triangular faces.

c a rectangular base.

d two circular faces.

1 Draw and describe the following 3D shapes:

a square pyramid

b cone

c hexagonal pyramid

d cylinder

- -

2 Choose two of the 3D shapes in question 1 and compare them on a Venn diagram.

Object 1: _____

Object 2: _____

OXFORD UNIVERSITY

1. Zak made a 3D wooden shape and painted it as a gift for his mother. Draw what the object might have looked like and the pieces of wood he needed to make it.

2. Georgie made a sculpture with five different 3D shapes. Draw what it might have looked like and label each of the 3D shapes.

Practice

1 Record how many of each angle type the shapes have.

a

right angles ☐

smaller than a right angle ☐

larger than a right angle ☐

b

right angles ☐

smaller than a right angle ☐

larger than a right angle ☐

c

right angles ☐

smaller than a right angle ☐

larger than a right angle ☐

d

right angles ☐

smaller than a right angle ☐

larger than a right angle ☐

---

2 Draw the other half of the crocodile's mouth so that it makes:

a a right angle.

b smaller than a right angle.

c larger than a right angle.

1

a Find as many right angles in the picture as you can and trace over them in red.

b Find as many angles as you can that are smaller than a right angle and trace over them in blue.

c Find as many angles as you can that are larger than a right angle and trace over them in green.

---

2 Draw your own picture with at least:

a four right angles.

b three angles smaller than a right angle.

c two angles larger than a right angle.

Use the same colours to trace the angle types as in question 1.

1 Ella drew a shape with four right angles. What might it have looked like? Show at least two examples.

2 Layton drew a shape with one right angle. What might it have looked like? Show at least two examples.

3 Look around you and find examples of right angles and angles that are smaller and larger than right angles. Draw and label examples of each.

OXFORD UNIVERSITY

Practice

1 Find and circle six items with line symmetry in the picture. Draw the lines of symmetry on the items.

2 Draw the other half of each shape so it is symmetrical.

a

b

c

d

e

f

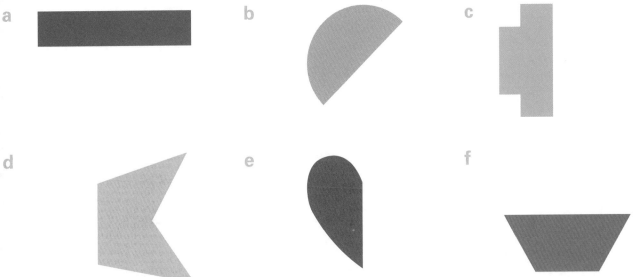

1

a Write your name in capital letters.

_____

b List the capital letters in your name that have line symmetry and draw the lines in.

_____

2

a Write your name in lower case letters.

_____

b List the lower case letters in your name that have line symmetry and draw the lines in.

_____

3 Draw your own picture with at least six items that have line symmetry. Draw the lines of symmetry on each item.

OXFORD UNIVERSITY

1   Make a pattern on the grid below that has both horizontal and vertical line symmetry.

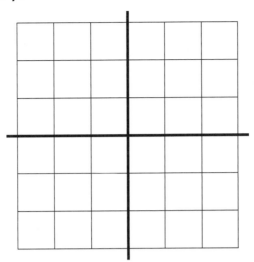

2

a   List some words where every letter has line symmetry.

b   List some words where none of the letters have line symmetry.

Practice

1 Redraw each shape after a slide, a flip and a turn.

a

| slide | flip | turn |

b

| slide | flip | turn |

c

| slide | flip | turn |

d

| slide | flip | turn |

e

| slide | flip | turn |

OXFORD UNIVERSITY

1 Draw a shape that:

a looks the same when it is flipped horizontally.

b looks different when it is flipped horizontally.

c looks the same after a quarter turn clockwise.

d looks different after a quarter turn clockwise.

e looks the same after a horizontal flip, but different after a quarter turn clockwise.

f looks different after a horizontal flip, but the same after a quarter turn anticlockwise.

g looks different after both a horizontal flip and a quarter turn clockwise.

h looks different after both a horizontal and a vertical flip.

1 Design a pattern that has:
- at least one horizontal flip
- at least one vertical flip
- at least one quarter turn
- at least one slide.

2 Describe how you used flips, slides and turns in your pattern.

OXFORD UNIVERSIT

Practice

A B C D E F G H I

1 Write the location of:

a the green tent. _____

b the barbeque. _____

c the tallest tree. _____

d the caravan. _____

e the picnic table. _____

f the playground. _____

2 Draw:

a a shower block in H6 and I6.

b a campfire in B3.

c a river in A5, B5 and B6.

d a car park in F1 and G1.

What other features might you add to the map?

3 Write directions to get from the playground to the car.

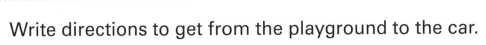

Alientown

1 Add and label the following features on the map of Alientown:

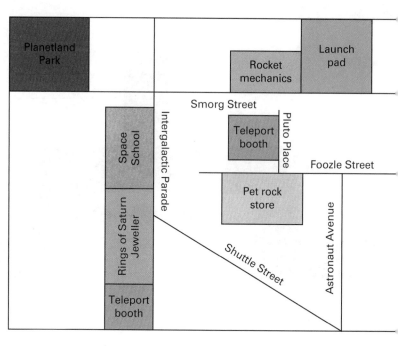

a a new road from Smorg Street near Planetland Park, parallel to Intergalactic Parade.

b a food shop on Shuttle Street.

c a slime factory at the corner of Foozle Street and Astronaut Avenue.

d a robot repair shop on Intergalactic Parade.

2 Write directions from one teleport booth to the other if you had to walk instead of teleporting.

3 Imagine you were standing at the corner of Shuttle Street and Intergalactic Parade. Describe what you see and where you might go.

OXFORD UNIVERSITY

1 Create a map of your ideal town. Include the following:

- roads
- shops
- a hospital
- a school

- houses or other places to live
- other places you would enjoy going to
- boundaries between different areas of your map.

2 Write clues to help someone find two different features on your map.

3 Write directions from the school to the hospital.

Practice

1 List a data source you could use to find out:

a how many of her spelling words Portia knew. _____

b how many red cars are in the car park. _____

c how Year 3 students feel about mathematics. _____

d how many people in your home country
speak more than one language. _____

e the favourite TV shows of nine year olds. _____

2

a List the first letter of the first name of each student in your class.

b Record the data you collected in question 2a using tally marks.

First letter	Number of people

Surveys and observation are the most common data sources.

OXFORD UNIVERSITY

1

a   Predict the answers you might get to the following survey question:
    What did you have for breakfast today?

_____

_____

b   Survey at least 10 people and record their responses in a list.

c   Use tally marks to organise the data.

List   Tally marks

Breakfast	Tally

d   Write two statements about whether the data was as you predicted.

2   List two other survey questions you could ask people in your class.

1 Eric collected data about his fellow students' favourite subjects.

a What source might he have used? _____

b Construct a table to show what the results might have looked like if there were 25 people in his class.

2 Caitlin conducted a survey. Her results are shown below. What question might she have asked? Give at least two options.

Colour	Tally
Red	ⵌ ⵌ ||||
Blue	ⵌ ||||
Green	||
Black	ⵌ |
Other	||||

OXFORD UNIVERSITY

1 Label the following parts of the graph:

a   the title   b   the *x*-axis   c   the *y*-axis

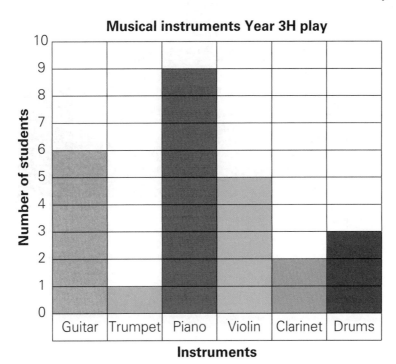

2 Use the graph to find:

a   how many different instruments are played. _____

b   the highest number on the *y*-axis. _____

c   the number of students who play the most popular instrument.

_____

d   the least popular instrument. _____

e   the difference between the number of people who play piano and

those who play violin. _____

f   the total number of answers recorded. _____

1 Use the data in the bar graph to make a pictograph.

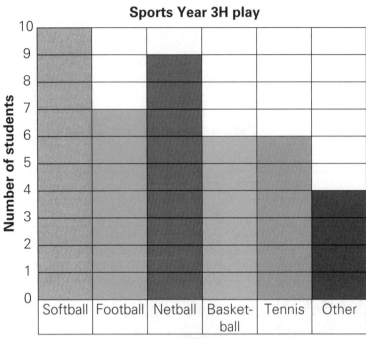

2

a Which sport is the most popular? _____

b Which two sports do the same number of people play?

_____

c What do you think the source of the data is? _____

OXFORD UNIVERSITY

1 Below is a list of responses people gave when asked what their favourite kinds of books are.

action	romance	adventure	action	non-fiction
humour	action	romance	humour	sad
adventure	adventure	sad	action	romance
non-fiction	non-fiction	non-fiction	action	humour

a Display the data in two different ways.

b How are your two displays similar? _____

_____

c How are they different? _____

_____

Practice

1 Use the data to answer the questions.

Vegetables in the school garden

			🥒	
			🥒	
			🥒	🥕
		🫑	🥒	🥕
🥦		🫑	🥒	🥕
🥦		🫑	🥒	🥕
🥦	🥬	🫑	🥒	🥕
🥦	🥬	🫑	🥒	🥕
Broccoli	Cauliflower	Capsicum	Zucchini	Carrot

a Which two vegetables did the school grow most of?

_____

_____

b How many more capsicums were there than cauliflowers?

_____

c How many vegetables were grown in total?

_____

2 Write three of your own questions that can be answered by the graph above.

a _____

b _____

c _____

3 Write the answers to your questions.

a _____

b _____

c _____

How is a pictograph different from a bar graph?

OXFORD UNIVERSITY

These two graphs show the favourite lunchtime activities of two different classes.

Favourite lunchtime activities of 3F

Number of students

10					
9					
8					
7					
6					
5					
4					
3					
2					
1					
0	Reading	Games	Talking	Eating	Sport

Activities

Favourite lunchtime activities of 3P

| Reading | Games | Talking | Eating | Sport |

a Write two statements about how the data on the two graphs is similar.

b Write two statements about how the data on the two graphs is different.

c How are the graphs similar and different?

1 The Year 3 students at Mathematics School were surveyed about their favourite desserts. There were five different answers: apple pie, ice-cream, cheesecake, chocolate, cake. Ice-cream was the most popular. Apple pie was the least popular.

a Draw a graph to show what the data might look like.

b Write three questions about the data.

c Write three statements about the data.

OXFORD UNIVERSIT

Practice

1) Use the Venn diagram showing sports played by some Year 3 students to answer the questions.

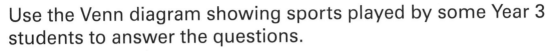

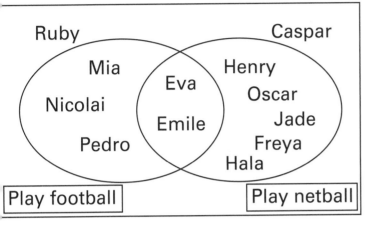

Ruby

Caspar

Mia

Eva

Henry

Nicolai

Oscar

Emile

Jade

Pedro

Freya

Hala

| Play football | | Play netball |

a Which of the two sports do more people play?

_____

b How many people don't play either sport? _____

c How many people are represented on the diagram in total? _____

d Write two other statements using the information in the Venn diagram.

_____

_____

2) Use the Carroll diagram showing the preferred ice-cream flavours of some Year 3 students to answer the questions.

	Chocolate	Other
Boys	Layton Leo Morten Tsai Jacob Chan	Evan Ryan Dylan
Girls	Sophia Maya	Olivia Steph Amy Jing

a How many people are represented on the diagram in total? _____

b How many people prefer chocolate ice-cream? _____

c Which flavour category is more popular with girls?

_____

d Write two other statements using the information in the Carroll diagram.

_____

_____

1 The Venn diagram below has three circles. Some overlapping sections show where items in two circles are common. The middle section is where all three numbers match the criteria.

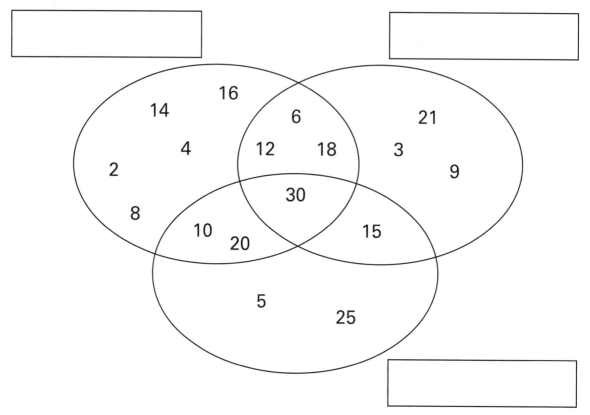

a Write labels in the boxes to show how the numbers were sorted.

b Sort the following numbers into the diagram: 60, 22, 24, 35, 27, 45.

c List three numbers that do not fit in any of the circles. _____

---

2 This Carroll diagram shows words sorted into categories.

a Write labels for each of the categories on the diagram.

b Add two words to each section.

	cat big lip	flat blue most
	arm inn urn	iris able axis

1 Edison compared 20 different animals on a Venn diagram. Show what his diagram might have looked like.

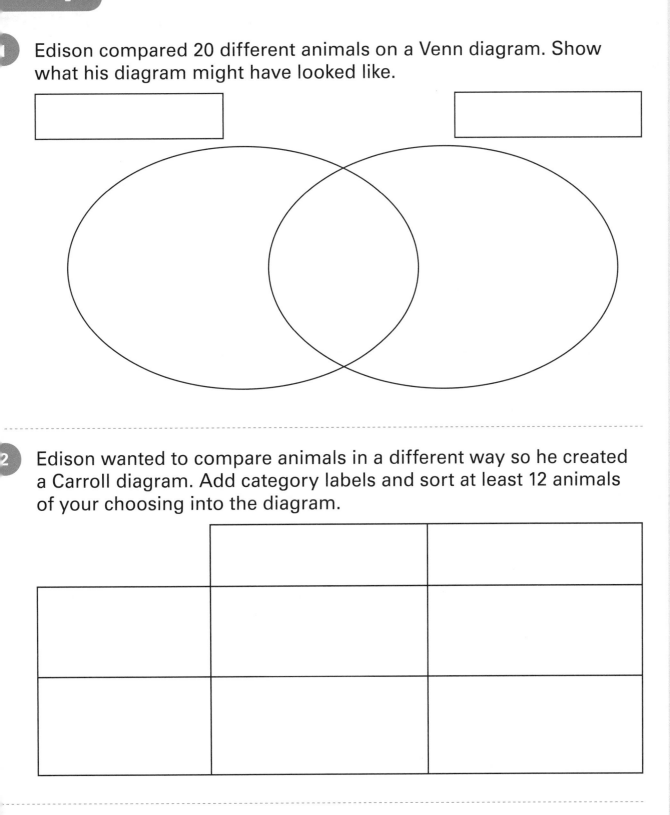

2 Edison wanted to compare animals in a different way so he created a Carroll diagram. Add category labels and sort at least 12 animals of your choosing into the diagram.

3 Monique went to a cafe that had options for two drinks, three sandwiches and three cakes. On another piece of paper, create a tree diagram to show all the different meals she might have had.

Practice

1 Dustin's restaurant serves three different main meals and three different desserts.

a Predict how many different combinations of mains and desserts you could have. _____

b The dishes Dustin serves are:

Mains – steak, chicken, vegetarian

Desserts – pie, ice-cream, muffin

Record and count all the possible meal combinations.

c Dustin adds lamb to his mains list and cake to his desserts. Record and count all the combinations that are now possible.

d What do you notice about the number of outcomes in each question?

OXFORD UNIVERSITY

1

a   Colour the spinner so that it has six possible outcomes.

b   Describe the likelihood of each of your outcomes being spun.

_____

_____

_____

_____

2   Ben had enough money to buy eight pieces of clothing. Could he make more different outfits if he bought two pairs of pants and six tops, or if he bought four of each? Show the options and explain your conclusion.

1 Describe an event with:

a exactly two possible outcomes.

b exactly three possible outcomes.

c more than 10 possible outcomes.

2 Fiona was having a party. She bought a giant bag of sweets to share into party bags. If each party bag was likely to have a red sweet, unlikely to have a blue sweet and it was possible to have green and orange sweets, draw what each bag might look like.

OXFORD UNIVERSITY

Practice

1

a If you draw 20 playing cards from a standard deck, how many do you think will be red and how many black?

My prediction	Red	Black

b Conduct the experiment and record the outcomes.

Outcomes	Red	Black

c Describe how the results compared with your prediction.

2

a This time, predict how many of each suit you will draw in 20 trials.

	Hearts ♥	Diamonds ♦	Spades ♠	Clubs ♣
My prediction				

b Conduct the experiment and record the outcomes.

	Hearts ♥	Diamonds ♦	Spades ♠	Clubs ♣
Outcomes				

c Describe how the results compared with your prediction.

Challenge

1 Put six blue, one green and three red counters in a bag. Describe the chance of drawing out:

a a blue counter. _____

b a green counter. _____

c a yellow counter. _____

d a red counter. _____

2

a Predict the outcomes if you conduct 20 trials, returning the counters to the bag after each trial.

	Blue	Green	Yellow	Red
My prediction				

b Conduct the trials and record the actual outcomes.

	Blue	Green	Yellow	Red
Outcomes				

c Make a data display of the results.

d Describe whether the outcomes match your predicted probabilities.

OXFORD UNIVERSIT

1 Design a simple chance experiment where each outcome has the same chance of happening.

a Describe your experiment.

b List the possible outcomes and predict the outcomes over 20 trials.

c Conduct your experiment and record the outcomes.

d Write about whether or not the outcomes were as you expected.

UNIT 1: Topic 1

Practice

1. a Six thousand, four hundred and seven

   b Five thousand and thirty

   c Nine thousand, six hundred and nineteen

2. a 3572   b 1605   c 8084

3. a 2873 = 2000 + 800 + 70 + 3

   b 7091 = 7000 + 0 + 90 + 1 OR 7000 + 90 + 1

   c 4009 = 4000 + 0 + 0 + 9 OR 4000 + 9

   d 9797 = 9000 + 700 + 90 + 7

4. a 3652   b 8043

   c 5307   d 6198

Challenge

1. a 7 hundreds   b 5 tens

   c 3 hundreds   d 4 thousands

2. a 500   b 90   c 40

   d 7000   e 4   f 900

3. 1902, 2095, 7194, 7580, 7851, 9143

4. a

Th	H	T	O
1	9	4	8
1	9	9	8
3	5	2	0
6	7	9	6
6	8	8	1

   b

Th	H	T	O
4	0	2	9
4	1	2	8
4	1	3	4
4	9	5	1
5	3	0	7

Mastery

1. A range of answers is possible. Look for students who are able to flexibly rename numbers and who use a systematic approach. Likely answers include:

   8000 + 300 + 20 + 9

   8000 + 300 + 29

   8000 + 329

   8300 + 20 + 9

   8300 + 29

   8320 + 9

2. a A range of answers is possible. Look for students

with a systematic approach to choosing numbers – e.g.

2345	2354	2453
2435	3245	3254
3452	3425, etc.	

   b 3245   3254   c 3254

3. Multiple answers are possible. Look for students who can apply their knowledge of place value to generate a range of options – e.g.

Will	Max	Will	Max
9234	6571	8234	6271

Will	Max	Will	Max
1289	1073	4231	3970

UNIT 1: Topic 2

Practice

1. a odd   b even
   c odd   d odd
   e odd   f even
   g even   h odd
   i even

2. a

373	376	379	382	385
odd	even	odd	even	odd

388	391	394	397
even	odd	even	odd

   b

1000	1005	1010	1015	1020
even	odd	even	odd	even

1025	1030	1035	1040
odd	even	odd	even

   c

20	18	23	21	26
even	even	odd	odd	even

24	29	27	32	30
even	odd	odd	even	even

3.

Number	+7	Odd or even?
74	81	odd
391	398	even
2150	2157	odd
6708	6715	odd
7463	7470	even

Challenge

1. Predictions may vary. Correct responses are shown below.

   a even $2862
   b odd   8837 km
   c odd   4975 people
   d even   12 036 grains
   e odd   8229 people

Mastery

1. Multiple answers possible. Each number must end in a 6 or a 4.

2. Multiple answers possible. The answer given must be an odd number. Possible examples include:

   251   275   307   329   331   3

3. Multiple answers possible. Example given for each problem

   a 7821   b 2187
   c 7182   d 7218

4. a 97 621   b 97 612
   c 12 679   d 12 796

UNIT 1: Topic 3

Practice

1. a rearranging numbers to make them easier to add

   b extending number facts

   c extending doubles facts

   d split strategy

2. Strategies used will vary. You m like to share the strategies chos and the reasons students chose them as a class.

   a 40   b 68
   c 80   d 49

Challenge

1. a–b Students' own answers. Th most likely response is bungee jumping and helicopter ride at a total cost of $500.

2. a–b Students' own answers. Lo for students who have chosen numbers that suit the strategy and who can accurately total the amounts.

3. a–b Students' own answers. Lo for students who have chosen numbers that suit the strategy and who can accurately total the amounts.

4. a $254   b $378
   c $652   d $425

Mastery

1. Multiple answers possible. Sam answers are:

   90 + 4 = 94   32 + 62 = 9
   48 + 46 = 94   57 + 37 = 9
   19 + 75 = 94   1 + 93 = 94

a 64   **b** 90   **c** 54
d 108   **e** 96   **f** 132

Multiple answers possible. Teacher to check students' estimates for reasonableness. Sample answers are:

a 27 + 23 + 25 + 15 = 90
b 19 + 31 + 14 + 26 = 90
c 45 + 25 + 28 + 42 = 140
d 33 + 37 + 17 + 23 = 110

UNIT 1: Topic 4

Practice

a 79   **b** 868
c 968   **d** 893

Challenge

a 699   **b** 988
a 939   **b** 749
c Daria's   **d** Daria's

Mastery

Multiple answers possible. Look for students who recognise that the amounts must be above $150, and who can accurately calculate solutions.

Multiple answers possible. Look for students who are able to accurately add three numbers, and who demonstrate sound reasoning in their choice of numbers.

Multiple answers possible. Sample answers are: 321 + 456 = 777, 631 + 245 = 876 and 346 + 512 = 858.

Multiple answers possible. Sample answers are: 2345 + 6789 = 9134, 5793 + 2468 = 8261 and 4962 + 3875 = 8837.

UNIT 1: Topic 5

Practice

a 49
b 66
c 132

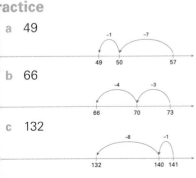

---

2 a 122   Split strategy
   b 341   Split strategy
   c 53   Extending number facts
   d 12   Split strategy
   e 163   Extending number facts

Challenge

1 a 320   **b** 338   **c** 234
2 a 82   **b** 203   **c** 121

3

Year level	Number of students	Number absent	Number present
F	37	9	28
1	42	6	36
2	34	8	26
3	73	5	68
4	66	9	57
5	41	4	37
6	45	7	38

Mastery

1. Multiple answers possible. Look for students who are able to use systematic thinking to generate answers. Sample answers are:

   174 − 43 = 131, 174 − 44 = 130, 174 − 45 = 129, 174 − 100 = 74, 174 − 74 = 100

2. Multiple answers possible. Look for students who can accurately subtract two different numbers from their selected totals.

3. 100 − 46 = 54 and 100 − 53 = 47

4. Multiple answers possible. Sample answers are: 321, 312, 456 and 431.

UNIT 1: Topic 6

Practice

1 a 53   **b** 423
   c 403   **d** 522

Challenge

1 a $412   **b** $410
2 a 586   **b** 202
   c 63   **d** 384

Mastery

1. Multiple answers possible. Look for students who are able to use reasoning and written subtraction methods to generate answers within the given range. Sample answers are:

   2574 − 924 = 1650
   2574 − 875 = 1699
   2574 − 969 = 1605

---

2. Multiple answers possible. Look for students who are able to interpret the problem and correctly generate both parts of the answers. Sample answers are:

   836 − 100 − 639 = 97
   836 − 36 − 703 = 97
   836 − 412 − 327 = 97

3. Multiple answers possible. Look for students who are able to work systematically to identify both the subtrahend and the answer, and who can work with numbers of different sizes. Sample answers are:

   557 − 100 = 457
   4500 − 4043 = 457
   9999 − 9542 = 457

UNIT 1: Topic 7

Practice

1 a 60 − 26 = 34, 60 − 34 = 26
   b 125 − 58 = 67, 125 − 67 = 58
   c 471 − 162 = 309,
      471 − 309 = 162
   d 998 − 853 = 145,
      998 − 145 = 853

2 a 49 + 18 = 67, 18 + 49 = 67
   b 28 + 65 = 93, 65 + 28 = 93
   c 97 + 289 = 386
      289 + 97 = 386
   d 673 + 131 = 804
      131 + 673 = 804

3 529 + 348 = 877
   348 + 529 = 877
   877 − 348 = 529
   877 − 529 = 348

Challenge

1 a 132 − 71 = 61
     71 + 61 = 132 OR 61 + 71 = 132
   b 5 × 9 = 45
     45 ÷ 9 = 5 OR 45 ÷ 5 = 9
   c 257 + 124 = 381
     381 − 257 = 124
     OR 381 − 124 = 257
   d 36 ÷ 9 = 4
     4 × 9 = 36 OR 9 × 4 = 36

2 a 46 + 30 − 1 = 75 km
   b 323 + 40 − 2 = 361

Mastery

1 a Multiple answers possible. Look for students who can use numbers flexibly and systematically to find solutions. Sample answers are:

$26 + 32 = 58, 32 + 26 = 58,$
$58 - 26 = 32, 58 - 32 = 26$

$56 + 33 = 89, 33 + 56 = 89,$
$89 - 56 = 33, 89 - 33 = 56$

$66 + 31 = 97, 31 + 66 = 97,$
$97 - 66 = 31, 97 - 31 = 66$

 b Multiple answers possible. Look for students who can use numbers flexibly and systematically to find solutions. Sample answers are:

$186 - 36 = 150, 186 - 150 = 36,$
$36 + 150 = 186, 150 + 36 = 186$

$683 - 76 = 607, 683 - 607 = 76,$
$76 + 607 = 683, 607 + 76 = 683$

$387 - 16 = 371, 387 - 371 = 16,$
$16 + 371 = 387, 371 + 16 = 387$

2 Multiple answers possible. Look for students who are able to use their knowledge of inverse operations and who use a range of number sizes to generate answers. Sample answers are:

Started with 1, ended with 14: $1 + 30 - 17 = 14$

Started with 30, ended with 43: $30 + 30 - 17 = 43$

Started with 153, ended with 166: $153 + 30 - 17 = 166$

3 Multiple answers possible. Look for students who are able to use their knowledge of inverse operations and who use a range of number sizes to generate answers. Sample answers are:

Started with 10, ended with 41: $10 \times 2 + 21 = 41$

Started with 87, ended with 195: $87 \times 2 + 21 = 195$

Started with 200, ended with 421: $200 \times 2 + 21 = 421$

UNIT 1: Topic 8

Practice

1 a $56 \div 7 = 8, 56 \div 8 = 7$
 b $36 \div 9 = 4, 36 \div 4 = 9$
 c $500 \div 100 = 5, 500 \div 5 = 100$
 d $84 \div 7 = 12, 84 \div 12 = 7$

---

2 a $7 \times 4 = 28, 4 \times 7 = 28$
 b $9 \times 6 = 54, 6 \times 9 = 54$
 c $11 \times 9 = 99, 9 \times 11 = 99$
 d $100 \times 4 = 400, 4 \times 100 = 400$

3 $10 \times 25 = 250, 25 \times 10 = 250,$
$250 \div 10 = 25, 250 \div 25 = 10$

Challenge

1 a $5 \times 8 = 40$ OR $8 \times 5 = 40$
$40 \div 8 = 5$ OR $40 \div 5 = 8$

 b $6 \times 7 = 42$ OR $7 \times 6 = 42$
$42 \div 6 = 7$ OR $42 \div 7 = 6$

 c $70 \div 7 = 10$
$7 \times 10 = 70$ OR $10 \times 7 = 70$

 d $63 \div 9 = 7$
$7 \times 9 = 63$ OR $9 \times 7 = 63$

2 a $4 \times 13 = 52, 52 \div 13 = 4,$
$52 \div 4 = 13, 13 \times 4 = 52$

 b $17 \times 15 = 255, 15 \times 17 = 255,$
$255 \div 17 = 15, 255 \div 15 = 17$

Mastery

1 Multiple answers possible. Sample answers include:

$4 \times 8 = 32$	$32 \div 8 = 4$
$8 \times 4 = 32$	$32 \div 4 = 8$
$3 \times 4 = 12$	$12 \div 3 = 4$
$4 \times 3 = 12$	$12 \div 4 = 3$
$14 \times 2 = 28$	$28 \div 14 = 2$
$2 \times 14 = 28$	$28 \div 2 = 14$

2 30, 45, 60, 75, 90

3 Numbers can be in any order. Possible answers are:

$1 \times 48, 2 \times 24, 3 \times 16, 4 \times 12, 6 \times 8$

UNIT 1: Topic 9

Practice

1 a halve and halve again
 b skip counting
 c using multiplication facts for division
 d double and double again

2 Students' own answer for strategies. Look for students who are able to accurately use their chosen strategies.
  a 21   b 40   c 100
  d 9   e 9   f 120

Challenge

1 a $64   b 20 crackers
  c $48   d 80 km   e $160

---

2 a $11 \times 3 = 11 \times 2 + 11 = 22 +$
$= 33$

 b $23 \times 3 = 23 \times 2 + 23 = 46 +$
$= 69$

 c $70 \times 3 = 70 \times 2 + 70 = 140 +$
$70 = 210$

 d $120 \times 3 = 120 \times 2 + 120 = 24$
$+ 120 = 360$

 e $36 \times 3 = 36 \times 2 + 36 = 72 +$
$= 108$

Mastery

1 Multiple answers possible. Look for students who understand that the number must be divisible by 3, who show the ability to work with numbers of a range of sizes and who demonstrate an understanding of a plausible response to the scenario.

2 Multiple answers possible. Sample answers are:

Jayda has $4; her brother has $1

Jayda has $20; her brother has $

Jayda has $32; her brother has $

3 a–c Answers will vary. Look for students who are able to choose realistic amounts for their chosen foods and who can accurately calculate both the totals and the shares.

 d Possible equations are:

$736 \div 4 = 184$ $776 \div 4 = 19$

UNIT 1: Topic 10

Practice

1 a $4 \times 37 = 4 \times 30 + 4 \times 7$
$= 120 + 28$
$= 148$

×	30	7	
4	120	28	= 148

 b $5 \times 64 = 5 \times 60 + 5 \times 4$
$= 300 + 20$
$= 320$

×	60	4	
5	300	20	= 32

 c $8 \times 55 = 8 \times 50 + 8 \times 5$
$= 400 + 40$
$= 440$

×	50	5	
8	400	40	= 44

OXFORD UNIVERSITY

d $9 \times 32 = 9 \times 30 + 9 \times 2$
$= 270 + 18$
$= 288$

×	30	2	= 288
9	270	18	

e $6 \times 46 = 6 \times 40 + 6 \times 6$
$= 240 + 36$
$= 276$

×	40	6	= 276
6	240	36	

Challenge

a 168 eggs

b 390 muffins

a Mr Smith: 265 km, Mrs Burton: 355 km, Miss White: 110 km, Mrs Swain: 240 km, Mr Chappel: 195 km

b 710 km

c 156 km

Mastery

a $7 \times 28 = 196$ is the closest combination. Accept any other combinations that are close.

b $7 \times 56 = 392$

c Possible combinations are: $6 \times 37 = 222$, $7 \times 37 = 259$, $5 \times 56 = 280$

d $5 \times 37 = 185$

Multiple answers possible. Sample answers are: $5 \times 87 = 435$, $6 \times 75 = 450$, $5 \times 89 = 445$

Multiple answers possible. Sample answers are: $6 \times 98 = 588$, $7 \times 85 = 595$, $9 \times 56 = 504$

UNIT 1: Topic 11

Practice

a Answer is 29. Possible order is $14 + 6 + 9$ OR $9 + 6 + 14$ OR $6 + 9 + 14$ OR $9 + 14 + 6$ OR $6 + 14 + 9$.

b Answer is 70. Teacher to check order. Look for students who identify pairs of numbers that are easier to add in their reordering.

c Answer is 80. Teacher to check order. Look for students who identify pairs of numbers that are easier to add in their reordering.

d Answer is 42. Possible order is $7 \times 3 \times 2$ OR $3 \times 2 \times 7$ OR $3 \times 7 \times 2$ OR $2 \times 3 \times 7$ OR $2 \times 7 \times 3$.

e Answer is 90. Possible order is $2 \times 5 \times 9$ OR $5 \times 9 \times 2$ OR $5 \times 2 \times 9$ OR $9 \times 2 \times 5$ OR $9 \times 5 \times 2$.

f Answer is 240. Teacher to check order. Look for students who identify pairs of numbers that make the calculation easier in their reordering.

2 a $3 \times 4 = 12$

b
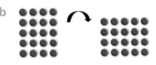
$4 \times 5 = 20$

c
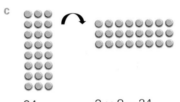
24        $3 \times 8 = 24$

d
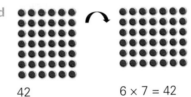
42        $6 \times 7 = 42$

Challenge

1 a 30   **b** 45

c 8   **d** 34

2 a $19 + 23 = 42$ $42 - 23 = 19$
$23 + 19 = 42$ $42 - 19 = 23$

b $8 \times 9 = 72$ $72 \div 8 = 9$
$9 \times 8 = 72$ $72 \div 9 = 8$

c $70 \div 10 = 7$ $7 \times 10 = 70$
$70 \div 7 = 10$ $10 \times 7 = 70$

d $84 - 36 = 48$ $48 + 36 = 84$
$84 - 48 = 36$ $36 + 48 = 84$

e $4 \times 8 \times 3 = 96$ $96 \div 3 \div 8 = 4$
Other possibilities are $4 \times 3 \times 8 = 96$, $3 \times 8 \times 4 = 96$, $3 \times 4 \times 8 = 96$, $8 \times 3 \times 4 = 96$, $8 \times 4 \times 3 = 96$, $96 \div 3 \div 4 = 8$, $96 \div 8 \div 3 = 4$, $96 \div 8 \div 4 = 3$, $96 \div 4 \div 3 = 8$, $96 \div 4 \div 8 = 3$

Mastery

1 Total number of fish is 125. Teacher to check combinations. Look for students who understand that they need to find pairs of numbers that add to 50 and pairs of numbers that add to 75, and who show systematic reasoning to find solutions.

2 Teacher to check. Multiple answers possible. Look for students who recognise combinations of numbers that will multiply to make 48, such as $12 \times 2 \times 2$ and $4 \times 4 \times 3$.

UNIT 2: Topic 1

Practice

1 a–f Teacher to check. Students may choose to colour any parts – look for students who colour the correct number as indicated in the questions.

2 Some variance is possible. Students are still correct if they have divided their shape into the correct number of parts and the parts are fairly even in size. The most likely divisions are shown below.

a
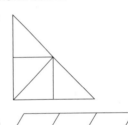

b
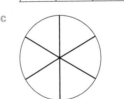

c
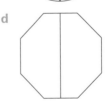

d
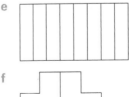

e
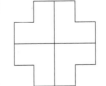

f

3 Teacher to check.

4 a $\frac{2}{4}$ b $\frac{2}{3}$ c $\frac{2}{6}$

d $\frac{2}{2}$ e $\frac{2}{8}$ f $\frac{2}{4}$

Challenge

1 **a–b** Answers will vary. Look for students who are able to accurately represent their colours as fractions.

2 **a–b** Answers will vary. Look for students who are able to accurately represent their colours as fractions.

3 Answers will vary. Some of the more creative possibilities are:

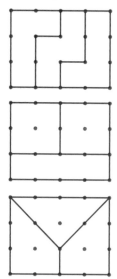

4 Answers will vary. Some of the more creative possibilities are:

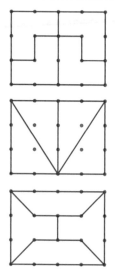

Mastery

1 a Multiple answers possible. Look for students who use numbers higher than 5 to show fifths and who can accurately draw the scenario.

b–d Answers will depend on the numbers chosen for the diagram. For example, if 10 crackers were drawn, that would be the number Gianni started with. He would have given away 6 crackers and kept 4 for himself.

e $\frac{2}{5}$

2 a Multiple answers possible. Look for students who use numbers higher than 4 to show quarters and who can accurately draw the scenario.

b–d Answers will depend on the numbers chosen for the diagram. For example, if 12 sweets were drawn, that would be the number Eleanor started with. She would have eaten 9 sweets and have 3 left.

e $\frac{3}{4}$

3 Diagrams will vary. Possible fraction options are:

Jonas	Sarah	Jonas	Sarah
$\frac{1}{6}$	$\frac{5}{6}$	$\frac{2}{6}$	$\frac{4}{6}$
Jonas	Sarah	Jonas	Sarah
$\frac{3}{6}$	$\frac{3}{6}$	$\frac{4}{6}$	$\frac{2}{6}$
Jonas	Sarah		
$\frac{5}{6}$	$\frac{1}{6}$		

UNIT 2: Topic 2

Practice

1 a

b

c

d

e

Challenge

1 Look for students who can evenly space the items on each number line. Labels on number lines should be:

a $\frac{1}{8}, \frac{2}{8}, \frac{3}{8}, \frac{4}{8}, \frac{5}{8}, \frac{6}{8}$

b $\frac{3}{5}, \frac{4}{5}, 1, 1\frac{1}{5}, 1\frac{2}{5}, 1\frac{3}{5}$

c $0, \frac{1}{3}, \frac{2}{3}, 1, \frac{4}{3}, \frac{5}{3}, 2$

d $0, \frac{1}{2}, 1, 1\frac{1}{2}, 2, 2\frac{1}{2}, 3, 3\frac{1}{2}$

e $0, \frac{1}{4}, \frac{2}{4}, \frac{3}{4}, 1, 1\frac{1}{4}, 1\frac{2}{4}, 1\frac{3}{4},$ $2, 2\frac{1}{4}, 2\frac{2}{4}, 2\frac{3}{4}$

2 a 8 b 7
c 11 d 6

Mastery

1 a

b $\frac{6}{8}, \frac{7}{8}, \frac{8}{8}$ or 1, $1\frac{1}{8}, 1\frac{2}{8}, 1\frac{3}{8}, 1\frac{4}{8}, 1\frac{5}{8},$ $1\frac{6}{8}, 1\frac{7}{8}, 1\frac{8}{8}$ or 2, $\frac{3}{4}, \frac{4}{4}$ or 1, $\frac{5}{4}, \frac{6}{4}, \frac{7}{4}, \frac{8}{4}$ or 2

c $\frac{1}{8}, \frac{2}{8}, \frac{1}{4}, \frac{3}{8}$

d $\frac{1}{4}$ & $\frac{2}{8}; \frac{2}{4}$ & $\frac{4}{8}; \frac{3}{4}$ & $\frac{6}{8}; \frac{5}{4}$ & $1\frac{2}{8}$ $\frac{6}{4}$ & $1\frac{4}{8}; \frac{7}{4}$ & $1\frac{6}{8}$

2 **a–c** Teacher to check that students have made a reasonabl estimate of where each fraction falls on the number line.

d Nala e Archer

UNIT 3: Topic 1

Practice

1 Teacher to check.

2 Teacher to check.

3 a E b A
c B d C

Challenge

1 **a–d** Answers will vary. Look for students who are able to accurately select coin combinations to meet the totals. Sample answers are:

a 5 × ¥5 OR 2 × ¥10 and 1 × ¥5

b 3 × ¥50 OR 1 × ¥100 and 1 × ¥50

c 3 × ¥100 OR 6 × ¥50

d 5 × ¥100 OR 10 × ¥50 OR 1 × ¥500

2 a 3 b 3
c 4 d 6

3 **a–d** Answers will vary. Look for students who are able to accurately select coin combinations to meet the totals. Sample answers are:

a ¥100 and ¥10

b 4 × ¥100, 1 × ¥50, 4 × ¥10 and 9 × ¥1

c 5 × ¥50

d 1 × ¥50 and 1 × ¥10

astery

Multiple answers possible. Look for students who recognise that the total of the two items must be $5.80 and who can use a systematic approach to identify different combinations.

Multiple answers possible. Look for students who are able to flexibly use the available denominations to generate different combinations.

Multiple answers possible. Look for students who are able to identify appropriate denominations with which Ruby might have paid and who can accurately calculate the cost of the item based on this amount.

NIT 4: Topic 1

actice

a Rule: Subtract 5

| | 65 | 60 | 55 | 50 | 45 | 40 | 35 | 30 | 25 |

b Rule: Add 2

| | 59 | 61 | 63 | 65 | 67 | 69 | 71 | 73 | 75 |

c Rule: Add 7

| | 24 | 31 | 38 | 45 | 52 | 59 | 66 | 73 | 80 |

d Rule: Subtract 10

| | 88 | 78 | 68 | 58 | 48 | 38 | 28 | 18 | 8 |

a 58, 70   b 72, 88, 96

c 50, 36, 8

hallenge

a Rule: Subtract 11

Incorrect number pair should be 45 and 34 OR 46 and 35

b Rule: Add 3

Incorrect number pair should be 98 and 101 OR 97 and 100

c Rule: Add 10

Incorrect number pair should be 245 and 255 OR 244 and 254

---

d Rule: Subtract 12

Incorrect number pair should be 72 and 60 OR 80 and 68

2 a Answers may vary. Most likely answer is odd numbers.

b Answers will vary. Look for students who recognise they need to identify odd numbers.

c Answers will vary. Look for students who recognise they need to identify even numbers.

Mastery

1 Multiple answers possible. Look for students who are able to use numbers of different sizes as starting points and who are able to accurately apply the rule to their chosen start numbers.

2 Multiple answers possible. Look for students who recognise appropriate combinations of operations and who are able to successfully apply the rules they identify.

3 Multiple answers possible. Look for students who choose an appropriate method to accurately represent their pattern and who can follow the criteria given in the problem.

UNIT 4: Topic 2

Practice

1 a 47 + 47 = 94 OR 47 × 2 = 94

b $143 – $37 – $24 = $82

c 213 + 157 = 370

d 56 + 28 – 40 = 44

2 a 13 + 11 = 24

b 10 = 18 – 8

c 27 – 14 = 13 OR 27 = 14 + 13

d 42 – 23 = 19 OR 42 = 23 + 19

Challenge

1 a 6   b 7

  c 23   d 11

2 a 317   b 60   c 64

  d Lizards, stick insects and parrots together

  e Cat and fish

  f Parrot

Mastery

1 a 7 + 6 + 7 = 20

b 25 – 6 + 2 = 21

c 25 – 10 – 4 = 11

d 58 + 12 + 4 = 74

e 105 + 7 – 8 + 12 = 116

f 132 – 5 – 6 + 3 = 124

2 Multiple answers possible. Look for students who are able to successfully add three numbers to make the given total and who can represent their equations as number sentences.

3 Multiple answers possible. Likely answers include: 24 bags of 12; 32 bags of 9; 36 bags of 8; 48 bags of 6; 72 bags of 4; 144 bags of 2; and the inverse of each of these.

UNIT 5: Topic 1

Practice

1 stop sign – 2 m, 50c coin – 3 cm, bottle – 30 cm, basketball court – 30 m, pencil – 15 cm

2 a Teacher to check. Look for students who can make a range of different shapes with the target area.

b 40 cm²

c 88 cm²

d Teacher to check. Answers will vary depending on the shapes drawn by students.

Challenge

1 NOTE: Allow a reasonable margin of error in the millimetre measurements.

a 5 cm or 52 mm

b 9 cm or 89 mm

2 a–d Teacher to check. Look for students who can accurately use a ruler, including where to position the zero point.

3

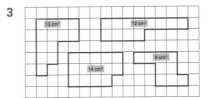

Mastery

1 **a–d** Teacher to check. Look for students who are able to accurately measure each of the designated items within a reasonable margin of error.

2 Teacher to check. Look for students who are able to identify appropriate items to measure and to measure them accurately.

3 Teacher to check. Look for students who are able to use tiles of the given areas to cover the floor without any gaps or overlaps.

UNIT 5: Topic 2

Practice

1 **a** Several different configurations possible. The most likely is:

Volume: 18 cm³

b Several different configurations possible. The most likely is:

Volume: 16 cm³

2 **a** 2 litres   **b** 9 litres

   c 200 mL   **d** 5 mL

   e 500 mL   **f** 4 litres

Challenge

1 **a** capacity   500 mL

   b capacity   50 litres

   c volume   32 cm³

2 **a–b** Teacher to check. Look for students who demonstrate a sound understanding of capacity in relation to a litre and who can draw on their own experiences to come up with examples.

Mastery

1 Teacher to check. Look for students who can successfully adhere to the criteria and who are able to draw a range of different configurations for their objects.

2 Multiple answers possible. Look for students who are able to successfully identify three capacities that make the given total and who show fluency in their calculations.

3 Multiple answers possible. Look for students who are able to successfully identify three capacities that make the given total and who show fluency in their calculations.

UNIT 5: Topic 3

Practice

1

Lighter than 1 kg	1 kg	Heavier than 1 kg
A	D	B
C		
E		
F		
G		
H		

2 **a** Possible combinations are: A & F, A & C, C & H, C & G, C & F, C & E, E & F, E & G, E & H, F & G, F & H, G & H

   b A & H

   c A & D, D & G

   d B and any of the other items

3 **a** 75 g   **b** 75 kg

   c 7.5 g   **d** 7.5 kg

Challenge

1 **a–d** Teacher to check. Look for students who can interpret the masses of items and place them on the correct ends of the scales according to their relative masses.

2 **a** 400 g

   b 1.5 kg OR 1500 g

   c 3 kg

Mastery

1 Multiple answers possible. Look for students who are able to both make reasonable estimates about the mass of different items and fluently work with masses to make a total of 3 kg.

2 **a** Multiple answers possible. Look for students who are able to make reasonable estimates about the mass of different animals.

b Answers will depend on answe to question 2a. Mark students correct if they have correctly ordered the masses, regardles of whether the estimates were reasonable or not.

UNIT 5: Topic 4

Practice

1 **a** 10 to 2   **b** 6:20

   c 5 past 8   **d** 20 to 1

   e 8:45   **f** 18 minutes past

2 **a** 2 minutes past 3   3:02

   b 7 minutes to 9   8:53

   c 29 minutes past 11   11:2

Challenge

1 **a**

   b

   c

2 **a** Thirty-four minutes past eight OR twenty-six minutes to nine

   b 8:52

   c 18 minutes

3 **a**

   b

OXFORD UNIVERSITY

c

d

e

f

Mastery

Teacher to check clocks and analogue times. Possible times are: 2:35, 2:53, 3:25, 3:52, 5:23, 5:32.

Multiple answers possible. Look for students who are able to successfully schedule the activities at logical times and represent the times accurately on an analogue clock, and who are able to make reasonable estimates of the duration of activities using appropriate units of time.

UNIT 6: Topic 1

Practice

Regular	Irregular
C	A
E	B
	D
	F

a–f Teacher to check. Look for students who can correctly interpret the criteria and draw a shape that meets the requirements.

Challenge

1 **a–d** Teacher to check descriptions. Look for students who show awareness of a range of attributes of shapes, such as angles, corners and parallel lines. Sample shapes are:

a

b

c

d

2 Teacher to check. Answers will depend on the shapes chosen. Look for students who can compare and contrast their shapes using the language of geometry.

Mastery

1 Multiple answers possible. Look for students who make a variety of shapes and are able to accurately name them.

2 Multiple answers possible. Look for students who recognise regular and irregular shapes and can use creativity and problem-solving skills to make their drawing according to the specifications.

UNIT 6: Topic 2

Practice

1

Prism	Pyramid	Other
A	D	B
C	F	E
G		
H		

2 **a–d** Teacher to check. Look for students who are able to draw a 3D shape that fits the given criteria. Likely shapes are:

a cube

b triangular prism

c rectangular prism

d cylinder

Challenge

1 **a–d** Teacher to check descriptions. Look for students who show awareness of a range of attributes of 3D shapes, such as faces, corners and edges. Sample shapes are:

a

b

c

d

2 Teacher to check. Answers will depend on the 3D shapes chosen. Look for students who can compare and contrast their shapes using the language of geometry.

Mastery

1 Teacher to check. Look for students who can successfully draw a 3D shape and identify the shapes of the faces.

2 Multiple answers possible. Look for students who are able to draw and identify a range of 3D shapes and creatively incorporate them in their design.

UNIT 7: Topic 1

Practice

1 **a** right angles 4

   smaller than a right angle 0

   larger than a right angle 0

   b right angles 0

   smaller than a right angle 0

   larger than a right angle 5

c	right angles	0
	smaller than a right angle	2
	larger than a right angle	2
d	right angles	2
	smaller than a right angle	2
	larger than a right angle	1

2 a

Note: b & c are sample answers. Check that the student draws an angle smaller than a right angle for b, and larger than a right angle for c.

b

c

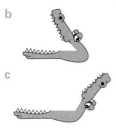

Challenge

1 **a–c** Teacher to check. A sample of each angle type is highlighted below.

2 **a–c** Teacher to check. Look for students who can consistently recognise angle types and use them in context.

Mastery

1 Teacher to check. The most likely responses are a square and a rectangle, but it is possible to have a shape with more than four sides.

2 Teacher to check. Many shapes are possible, including a right-angled triangle and an irregular pentagon or hexagon.

3 Teacher to check. Look for students who recognise angles in their environment and who can accurately classify them.

UNIT 8: Topic 1

Practice

1

2 a ▭ b ⬤

c ✚ d ✦

e ♥ f ⬡

Challenge

1 **a–b** Teacher to check. Answers will depend on the student's name and the formation of the letters.

2 **a–b** Teacher to check. Answers will depend on the student's name and the formation of the letters.

3 Teacher to check. Look for students who can identify objects in the environment that have line symmetry and can draw them with reasonable accuracy.

Mastery

1 Teacher to check. Look for students who are able to make a pattern that shows both horizontal and vertical symmetry.

2 a Multiple answers possible. It will also depend on how students form their letters and whether they use capital or lower case letters. Look for students who recognise both horizontal and vertical line symmetry in letters. Possible examples are: will, DAME, vow, BIT

b Multiple answers possible. It will also depend on how students form their letters and whether they use capital or lower case letters. Possible examples are: and, pet, pug, ram

---

UNIT 8: Topic 2

Practice

1 Answers may vary depending on the direction of the flips and turns. The most likely answers a shown below.

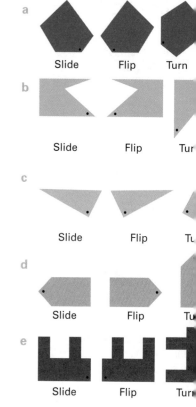

a Slide Flip Turn

b Slide Flip Tur

c Slide Flip Tu

d Slide Flip Tu

e Slide Flip Turn

Challenge

1 **a–h** Multiple answers possible. Teacher to check. Look for students who demonstrate a sound understanding of the three kinds of transformations and can apply this to identifying a shape that meets the criteria.

Mastery

1 Teacher to check. Look for students who are able to use their problem-solving skills and knowledge of transformations to make a pattern that meets the criteria.

2 Teacher to check. Look for students who can accurately use the language of transformations to describe their patterns.

ctice

B4	**b** E3
A6	**d** D2
F3	**f** I4

a–d

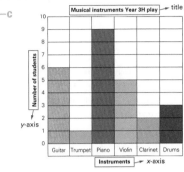

Teacher to check. Look for students who can use the language of direction to accurately describe the turns and direction of the path.

allenge

a–d Some variance possible. Teacher to check that placement meets the given instructions.

Teacher to check. Look for students who can identify the start and finish points and use the language of direction to formulate accurate instructions.

Teacher to check. Look for students who can use creativity and problem-solving skills as well as being able to accurately describe places or give directions based on the map.

stery

Teacher to check. Look for students who incorporate all the required features and show an awareness of map conventions.

Teacher to check. Look for students who can accurately use the language of direction to describe their chosen locations.

Teacher to check. Answer will depend on where students chose to place their schools and hospitals. Look for students who are able to draw on landmarks and street names to describe an accurate path between the location of the two places on their maps.

UNIT 9: Topic 1

Practice

1. The most likely responses are listed below. Allow other responses if students can offer a reasonable justification.

   a. test results

   b. observation

   c. survey

   d. Census

   e. survey

2. **a–b** Teacher to check. Look for students who can accurately record the data based on the students in your class.

Challenge

1. **a** Teacher to check. Look for students who make reasonable predictions about the types of responses they might receive.

   b–c Teacher to check. Look for students who can accurately use both data collection methods and check that the numbers in the two data sets match.

   d Teacher to check. Look for students who can compare their predictions with the results and draw some conclusions about any similarities and differences.

2. Teacher to check. Look for students who are able to generate questions on appropriate topics for the class and who can express their question in a way that suits collection of survey data.

Mastery

1. **a** The most likely response is survey. Accept other responses if students are able to offer a reasonable justification for their choice.

   b Answers will vary. Look for students who understand how to construct a table and who can choose appropriate categories for the data.

2. Multiple answers possible. Look for students who can suggest reasonable questions to match the data, such as what colour lunch box students have and what the favourite colour of particular students is.

UNIT 9: Topic 2

Practice

1. **a–c**

Musical instruments Year 3H play	← title

(bar graph: y-axis = Number of students 0–10; x-axis = Instruments: Guitar, Trumpet, Piano, Violin, Clarinet, Drums)

2. **a** 6   **b** 10
   c 9   **d** Trumpet
   e 4   **f** 26

Challenge

1. Teacher to check. Look for students who can accurately represent the data in a pictograph and who adhere to all the conventions of pictographs, such as titles and legends.

2. **a** Softball

   b Basketball and tennis

   c Teacher to check. Look for awareness of an appropriate data source such as a survey.

Mastery

1. **a** Teacher to check. Look for students who can choose appropriate methods such as pictographs, bar graphs or tables and who can accurately represent the data on each.

   b Teacher to check. Look for students who understand they are using the same data and who can identify elements of their displays that are similar, such as the visual effect of representing the larger and smaller data categories.

   c Teacher to check. Look for students who are able to compare the attributes of both displays, such as the way that data is represented and the way the scales work.

UNIT 9: Topic 3

Practice

1 a Zucchinis and carrots

   b 3

   c 25

2 a–c Teacher to check. Look for students who can formulate a range of questions drawing on the given data.

3 a–c Teacher to check. Look for students who can accurately answer the questions that they posed.

Challenge

1 a Teacher to check. Look for students who can make accurate comparisons between the data by finding similarities, such as the fact that the same number of students were surveyed in each class, sport was the most popular response in each class and seven students in each class preferred talking.

   b Teacher to check. Look for students who can make accurate comparisons between the data by finding differences, such as the fact that one class had games as the least popular activity while the other had eating, and that different numbers of students preferred activities such as reading and playing games.

   c Teacher to check. Look for students who can make accurate comparisons between the data types. Similarities include that both graphs show the same number of categories and the same titles of categories, and that both graphs have a title. Differences include that the bar graph shows the number of items in each category while you have to count the pictures in the pictograph.

Mastery

1 a Teacher to check. Look for students who choose an appropriate graph type for their data display and who can make plausible guesses about the number of students in each category based on a Year 3 cohort.

   b Teacher to check. Look for students who can formulate a range of questions that can be answered by the data.

   c Teacher to check. Look for students who can interpret the data accurately to formulate their statements.

UNIT 9: Topic 4

Practice

1 a Netball

   b 2

   c 12

   d Teacher to check. Look for students who can interpret the data to make accurate statements.

2 a 15

   b 8

   c Other

   d Teacher to check. Look for students who can interpret the data to make accurate statements.

Challenge

1 a–d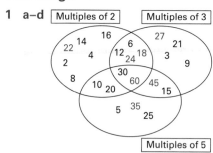

   c Any prime numbers will not fit into the categories, e.g. 7, 11, 13, 17, 19, 23, 29.

2 a Horizontal categories are 3- and 4-letter words. Vertical categories are words starting with consonants and words starting with vowels. Exact wording used by students may vary.

   b Teacher to check. Look for students who are able to understand the category requirements and choose words that accurately fit each.

Mastery

1 Teacher to check. Look for students who can identify appropriate categories and who can sort their chosen data accurately into the Venn diagram.

2 Teacher to check. Look for students who can identify appropriate categories and who can sort their chosen data accurately into the Carroll diagram.

3 Students may choose specific dish options. The diagram below represents the options with gen labels.

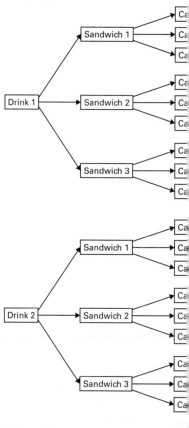

UNIT 10: Topic 1

Practice

1 a Teacher to check. Look for reasonable estimates based the variables in question.

   b There are 9 possibilities: steak and pie, steak and ice-cream, steak and muffin chicken and pie, chicken and ice-cream, chicken and muf vegetarian and pie, vegetarian and ice-cream, vegetarian a muffin

OXFORD UNIVERSIT

c There are 16 possibilities: steak and pie, steak and ice-cream, steak and muffin, steak and cake, chicken and pie, chicken and ice-cream, chicken and muffin, chicken and cake, vegetarian and pie, vegetarian and ice-cream, vegetarian and muffin, vegetarian and cake, lamb and pie, lamb and ice-cream, lamb and muffin, lamb and cake

d Teacher to check. Look for students who recognise patterns, such as the fact that if you multiply the number of options together it will give you the number of combinations.

allenge

a Teacher to check. Look for students who understand they need to use six different colours to have 6 possible outcomes.

b Teacher to check. Look for students who use the language of probability to accurately describe the chance of the colours on their spinner being spun.

There are 12 possible outcomes if Ben buys 2 pairs of pants and 6 tops, and 16 possible outcomes if he buys 4 of each. Exact combinations will vary depending on how students decide to describe them.

astery

a Answers will vary. Possibilities include tossing a single coin, or whether a baby will be a boy or a girl.

b Answers will vary. Possibilities include landing on a particular colour on a spinner with three sections, or drawing a triangle out of a bag by a particular corner.

c Answers will vary. Possibilities include who the next person to walk through the classroom door will be, or which student in the class will win a game such as bingo.

2 Multiple answers possible. Look for students who understand that not all the bags will look the same, and who can accurately interpret the criteria.

UNIT 10: Topic 2

Practice

1 a–b Teacher to check. Look for students who make plausible predictions and can accurately record the experiment data.

c Teacher to check. Look for students who recognise the likelihood of each outcome and demonstrate an understanding of why the results might differ from this.

2 a–b Teacher to check. Look for students who make plausible predictions and can accurately record the experiment data.

c Teacher to check. Look for students who recognise the likelihood of each outcome and demonstrate an understanding of why the results might differ from this.

Challenge

1 a–d Blue $\frac{6}{10}$ OR likely; green $\frac{1}{10}$ OR unlikely; yellow 0 OR impossible; red $\frac{3}{10}$ OR unlikely.

2 a Teacher to check. Look for students who are able to make reasonable predictions that match the likelihood words they chose in question 1.

b Teacher to check. Look for students who can accurately record the results of their trials.

c Teacher to check. Look for students who can choose an appropriate data display and who can accurately represent their data on it.

d Teacher to check. Look for students who understand why outcomes may be different from predicted probabilities.

Mastery

1 a–d Teacher to check. Look for students who show problem solving and reasoning skills in the design of their experiments and who can follow procedures to accurately record and analyse the outcomes of their experiment.